THE CITY

LONDON GOVERNMENT AND THE WELFARE SERVICES

INEQUALITY IN THE CITY

LONDON GOVERNMENT AND THE WELFARE SERVICES

S.K. RUCK

Routledge
Taylor & Francis Group
LONDON AND NEW YORK

First published in 1963

This edition published in 2007
Routledge
2 Park Square, Milton Park, Abingdon, Oxon, OX14 4RN
711 Third Avenue, New York, NY 10017

Routledge is an imprint of Taylor & Francis Group, an informa business

First issued in paperback 2013

Transferred to Digital Printing 2007

British Library Cataloguing in Publication Data
A CIP catalogue record for this book
is available from the British Library

London Government and the Welfare Services

ISBN13: 978-0-415-41765-5 (hbk volume)
ISBN13: 978-0-415-86473-2 (pbk volume)
ISBN13: 978-0-415-41926-0 (subset)
ISBN13: 978-0-415-41318-3 (set)

Routledge Library Editions: The City

THE GREATER LONDON GROUP

London Government and the Welfare Services

by

S. K. Ruck

LONDON
ROUTLEDGE & KEGAN PAUL

First published in 1963
by Routledge & Kegan Paul Limited
Broadway House, 68-74 Carter Lane
London, E.C.4

Printed in Great Britain
by Charles Birchall & Sons Ltd.
Liverpool & London

To

G. B. R.

Contents

PART THREE

Tables

Tables

Appendix

Foreword

by Professor W. A. Robson

THIS book contains the results of an inquiry carried out by the author in his capacity of a Research Officer for the Greater London Group at the London School of Economics and Political Science.

Mr. Ruck's qualifications for undertaking this work are exceptional. On leaving Oxford he lived for some years in tenement buildings in Bermondsey running boys clubs and working in the Borstal Service. In 1929 he became Secretary of the Second Survey of London Life and Labour carried out at the London School of Economics, and then served for many years in the Welfare Department of the London County Council, becoming its Deputy Chief Officer. In 1959–60 he held a Senior Simon Research Fellowship in the University of Manchester, which enabled him to study the effects of local government organisation on the Welfare Services for old people.

The research conducted by Mr. Ruck was made possible by a grant from the City Parochial Foundation. The members of the Group and the London School of Economics and Political Science are most grateful to the Trustees and Sir Donald Allen, the Secretary, for their generous help.

The Greater London Group consists at present of the following members:

W. A. Robson (Chairman), Professor Emeritus of Public Administration
A. C. L. Day, Reader in Economics
D. V. Donnison, Professor of Social Administration
R. C. Estall, Lecturer in Geography
S. J. Gould, Reader in Social Institutions
A. J. Greve, Lecturer in Social Administration
J. A. G. Griffith, Professor of English Law
Emrys Jones, Professor of Geography
J. E. Martin, Lecturer in Geography
E. J. Mishan, Lecturer in Economics
R. H. Pear, Reader in Political Science

G. J. Ponsonby, Sir Ernest Cassell Reader in Commerce (with special reference to Transport)
P. J. O. Self, Reader in Political Science
M. J. Wise, Professor of Geography

Research Staff:

S. K. Ruck, Research Officer
L. J. Sharpe, Research Officer
D. Peschek, Research Assistant

The Group was formed in 1958 in response to a letter from Sir Edwin Herbert, Chairman of the Royal Commission on Local Government in Greater London to the Vice-Chancellor of the University of London, expressing the hope that individual scholars or groups of scholars would be prepared to give disinterested evidence to the Royal Commission. The Group's initial objective was to conduct research and prepare evidence on the structure of London Government for the Royal Commission; and it was heavily occupied with this task during 1958 and 1959. The Nuffield Foundation provided most generous support for this and the subsequent work of the Group.

After completing their original task of preparing evidence for the Royal Commission the Greater London Group decided to embark on a further extensive programme of research into some of the major problems which will confront the newly created local authorities in Greater London.

Hitherto, the results of the Group's researches and its pronouncements (with the exception of its evidence before the Herbert Commission which was printed in the Minutes of Evidence of the Commission) have appeared in a series of Greater London Papers* published by the London School of

* The Greater London Papers are as follows:

No. 1	Education in Greater London	by A. V. Judges
No. 2	Theories of Local Government	by W. J. M. Mackenzie
No. 3	The Greater London Boroughs	by W. A. Robson
No. 4	Housing in Greater London	by J. B. Cullingworth
No. 5	Health, Welfare and Democracy in Greater London	by D. V. Donnison
No. 6	Transport in Greater London	by Ernest Davies
No. 7	Town Planning in Greater London	by Peter Self
No. 8	A Metropolis Votes	by L. J. Sharpe

Economics and Political Science. The length of Mr. Ruck's work, however, made it necessary for it to be published in book form.

Part I (Chapter I) outlines the scope of the study and offers some general comments on the existing administration. Part II (Chapters II–VI) comprises a detailed study of the administration of these services in the Greater London area by the local authorities concerned, on which the conclusions are based. The conclusions arrived at are embodied in Part III, (Chapter VII).

The appendix, which is properly an appendix to Chapter II, is a reproduction on a smaller scale in relation to South West Essex, of Miss K. M. Slack's study of services for the old in the County of London, published by the Codicote Press (7s. 6d.) under the title "Councils, Committees and Care for the Old".

The recommendations of the Royal Commission and the proposals of the Government White Paper relative to the Welfare Services were received with prophecies of disaster by numerous Aldermen, Councillors, politicians, welfare workers and magistrates of juvenile courts. Mr. Ruck's work suggests that these gloomy forebodings may be unfounded, and that the new structure will provide the London Boroughs with a valuable opportunity for reintegrating and reanimating the Welfare Services in the Metropolis.

The twofold merit of this study is, first, that it shows the extraordinary diversity in the quantity and quality in the provision of the Welfare Services in Greater London; and second, that it formulates a possible pattern of administration of these services by the new London boroughs which are being created in the Metropolis. Much of what the author proposes could with advantage be applied to county boroughs in other parts of the country. Thus the book has a value which extends far beyond the confines of the Metropolis.

London School of Economics
and Political Science
February, 1963

KEY TO MAP

Column A indicates the number of each new London Borough as shown on the map, Column B the former boroughs of which it is constituted with index letters referring to the map, Column C the populations of the former boroughs and the population of the new borough. C.B.—County Borough. M.B.—Metropolitan Borough. N.C.B.—Non-County Borough. U.D.—Urban District.

A	B	C
	a Paddington M.B.	115,000
	b St. Marylebone M.B.	69,000
	c City of Westminster M.B.	85,000
1		**269,000**
	a Hampstead M.B.	99,000
	b Holborn M.B.	22,000
	c St. Pancras M.B.	125,000
2		**246,000**
	a Finsbury M.B.	33,000
	b Islington M.B.	229,000
3		**262,000**
	a Hackney M.B.	165,000
	b Shoreditch M.B.	40,000
	c Stoke Newington M.B.	52,000
4		**257,000**
	a Bethnal Green M.B.	47,000
	b Popular M.B.	66,000
	c Stepney M.B.	92,000
5		**205,000**
	City of London	**5,000**
	a Greenwich M.B.	86,000
	b Woolwich M.B. (part)	144,000
6		**230,000**
	a Deptford M.B.	68,000
	b Lewisham M.B.	222,000
7		**290,000**
	a Bermondsey M.B.	52,000
	b Camberwell M.B.	175,000
	c Southwark M.B.	86,000
8		**313,000**
	Lambeth M.B.	223,000
	Wandsworth M.B. (part)	118,000
9		**341,000**
	a Battersea M.B.	106,000
	b Wandsworth M.B. (part)	229,000
10		**335,000**
	a Fulham M.B.	112,000
	b Hammersmith M.B.	110,000
11		**222,000**
	a Chelsea M.B.	47,000
	b Kensington M.B.	171,000
12		**218,000**
	a Chingford N.C.F.	46,000
	b Leyton N.C.B.	94,000
	c Walthamstow N.C.B.	109,000
13		**248,000**
	a Ilford N.C.B.	178,000
	b Wanstead & Woodford N.C.B.	61,000
	Chigwell U.D. (part)	7,000
	Dagenham N.C.B. (part)	2,000
14		**249,000**
	a Romford N.C.B.	115,000
	b Hornchurch U.D.	128,000
15		**243,000**
	a Barking N.C.B.	72,000
	b Dagenham N.C.B.	108,000
16		**179,000**
	a East Ham C.B.	105,000
	b West Ham C.B.	157,000
	Barking N.C.B. (part) } Woolwich M.B. (part) }	2,000
17		**264,000**
	a Bexley N.C.B.	90,000
	b Crayford U.D.	31,000
	c Erith N.C.B.	45,000
	Chislehurst & Sidcup U.D. (part)	44,000
18		**210,000**
	a Beckenham N.C.B.	77,000
	b Bromley N.C.B.	68,000
	c Chislehurst & Sidcup U.D. (part)	44,000
	d Orpington U.D.	80,000
	e Penge U.D.	26,000
19		**294,000**
	a Croydon C.B.	252,000
	b Coulsdon & Purley U.D.	75,000
20		**327,000**
	a Beddington & Wallington N.C.B.	33,000
	b Carshalton U.D.	57,000
	c Sutton & Cheam N.C.B.	79,000
21		**169,000**
	a Merton & Morden U.D.	68,000
	b Mitcham N.C.B.	64,000
	c Wimbledon N.C.B.	57,000
22		**189,000**
	a Kingston N.C.B.	36,000
	b Malden & Coombe N.C.B.	47,000
	c Surbiton N.C.B.	63,000
	Epsom & Ewell N.C.B. (part)	20,000
23		**166,000**
	a Barnes N.C.B.	40,000
	b Richmond N.C.B.	41,000
	c Twickenham N.C.B.	101,000
24		**182,000**
	a Brentford & Chiswick N.C.B.	55,000
	b Feltham U.D.	51,000
	c Heston & Isleworth N.C.B.	103,000
25		**209,000**
	a Hayes & Harlington U.D.	68,000
	b Ruislip & Northwood U.D.	73,000
	c Uxbridge N.C.B.	64,000
	d Yiewsley & West Drayton U.D.	24,000
26		**228,000**
	a Acton N.C.B.	65,000
	b Ealing N.C.B.	183,000
	c Southall N.C.B.	51,000
27		**300,000**
	a Wembley N.C.B.	125,000
	b Willesden N.C.B.	171,000
28		**296,000**
29	Harrow N.C.B.	**209,000**
	a Barnet U.D.	28,000
	b East Barnet U.D.	41,000
	c Friern Barnet U.D.	29,000
	d Finchley N.C.B.	69,000
	e Hendon N.C.B.	152,000
30		**318,000**
	a Hornsey N.C.B.	98,000
	b Tottenham N.C.B.	113,000
	c Wood Green N.C.B.	48,000
31		**259,000**
	a Edmonton N.C.B.	92,000
	b Enfield N.C.B.	110,000
	c Southgate N.C.B.	72,000
32		**274,000**

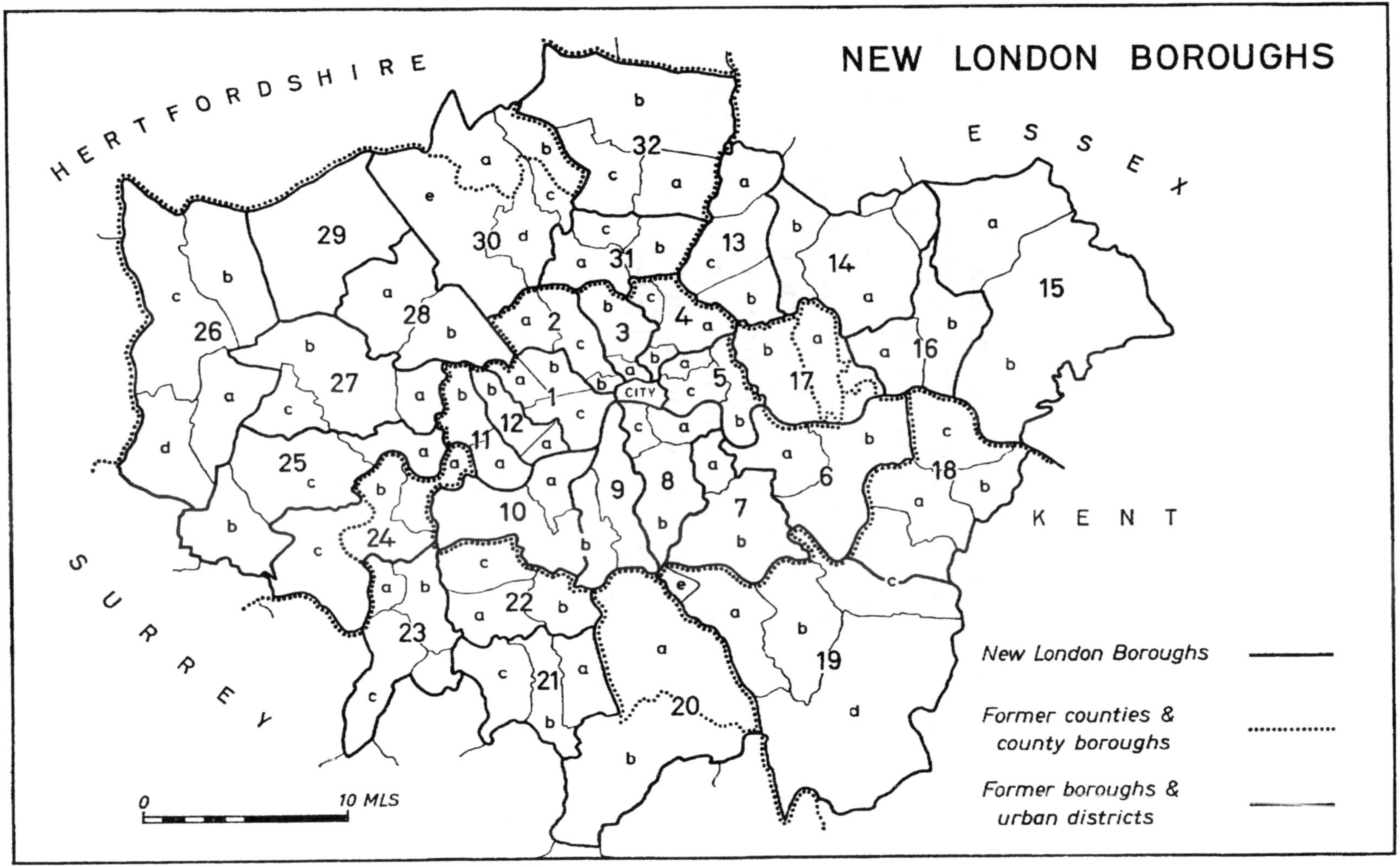
NEW LONDON BOROUGHS
HERTFORDSHIRE
ESSEX
KENT
SURREY
CITY
New London Boroughs
Former counties & county boroughs
Former boroughs & urban districts
0
10 MLS

ACKNOWLEDGEMENTS and thanks are due to the clerks of the various councils, medical officers of health, children's officers, chief welfare officers, housing managers and other officers of the local authorities and voluntary bodies, who gave so ungrudgingly of their time in supplying information and answering questions; to colleagues in the London School of Economics for criticism and advice, and in particular to the Drawing Office for producing the map; to the editor of the *Observer*, part of Chapter I having appeared in that journal.

S.K.R.

PART ONE

CHAPTER ONE

The Scope of the Services and the Responsible Authorities in Greater London

FOR the purposes of this study the Welfare Services are taken to be those which are provided by local authorities to give assistance, other than medical and financial, to those needing such assistance through the handicaps of age, or physical, mental or social infirmity. Their financial needs are taken care of by National Insurance and National Assistance, their medical needs by the National Health Service, all nationally administered except part of the Health Service.

Broadly speaking, the categories of those covered by the welfare services are the old, the physically handicapped (i.e. the blind, the deaf or dumb, and those substantially and permanently handicapped by illness, injury or congenital deformity), the mentally handicapped (i.e. the mentally retarded and the mentally ill discharged from hospital to community care) and the socially handicapped (children deprived of a normal home life, unmarried mothers, homeless and problem families and the "down and outs").

The chief statutes making provision for assistance to these groups are the National Assistance Act, 1948; the National Health Service Act, 1946 and the Mental Health Act, 1959; the Children's and Young Persons' Act, 1933 and the Children's Acts, 1948 and 1958. The main responsibility for the local implementation of these Acts is placed upon the first tier local authorities, i.e. the counties and county boroughs. The second tier authorities, i.e. the municipal boroughs, the urban and rural district councils and, in the County of London, the

Metropolitan Borough Councils, also have a part to play as Sanitary Authorities under the Public Health Acts. Furthermore under the National Assistance (Amendment) Act, 1962, they are empowered to provide meals and recreation for old people, and under Section 31 of the National Assistance Act, 1948, they may make contributions to the funds of any voluntary organisation whose activities consist of or include such provision. A wide construction has been placed upon this section by some second tier authorities, and extensive use made of it, as will be seen later.

Second tier authorities also come into the picture because they (and the county boroughs) are housing authorities while the counties are not,* and the housing of some of the "Welfare" categories mentioned above, especially the old, the crippled, and the so-called "Problem" families involves special considerations.

First tier authority functions in the area considered by the Herbert Commission† are undertaken by three county boroughs and six counties, two of them wholly within the area, the other four partly in the area and partly outside. The populations concerned were as follows:—

		Population in Review Area	*Population of Whole County*
County Boroughs:	Croydon	249,000	
	East Ham	109,900	
	West Ham	164,300	
Counties:	London	3,204,000	
	Middlesex	2,247,000	
	Essex	997,200	1,808,000
	Kent	541,130	1,666,000
	Surrey	904,940	1,446,000
	Hertfordshire	296,490	784,000

In the recommendations of the Herbert Report, the population of Hertfordshire to come into the Greater London Area was reduced to 98,880, and that of Kent to 497,170. Further reductions in the Government's final proposals brought the number transferred from Hertfordshire down to 69,000, from Surrey to 568,000 and from Essex to 924,000.

* The County of London is the one exception here.

† Royal Commission on Local Government in Greater London, Cmd. 1164, 1960.

The Royal Commission's recommendations envisaged 52 new Greater London Boroughs with populations (apart from the City) ranging from about 100,000 to 250,000. The Government's final proposals reduced this total to 32 (plus the City) with populations from 166,000 to 341,000. In this connection it is worth noting that of the 83 existing county boroughs in England and Wales, only 17 have a population of over 200,000 and only seven of over 300,000, so that the proposed new London boroughs will rank in size and resources with the largest of the existing county boroughs.

One of the main recommendations of the Herbert Report was that most local government services, including the welfare services, should be administered by these new London boroughs and so provide a more local, intimate and responsive service.

The three London county boroughs were all within the range of population suggested for the new London boroughs by the Royal Commission, and one, Croydon, within that proposed by the Government. It therefore seems a useful exercise to compare the administration of welfare services by the counties in the London area with that by the county boroughs, in order to consider whether or not these services are likely to suffer if and when they cease to be under county administration, as has been strongly maintained in the press and elsewhere by many associated with the health and children's services, especially in respect of the County of London.

No-one has yet devised a method of assessing the comparative merits of the various departments of different councils, but at least the amount a council is prepared to spend on its welfare services is some indication of the importance it attaches to them. Table I shows the amounts spent per 100 of the population on these services by the Welfare Authorities in the Greater London area in the latest available year.

The county boroughs, the administrative units most nearly comparable with the proposed new London boroughs, spend a markedly higher amount on the welfare services per head of the population and more on the children's services than the counties apart from London.

There is support here for the view that the welfare services

TABLE I.—EXPENDITURE ON WELFARE SERVICES, SOCIAL CLASS AND PROPORTIONS OF THE VERY OLD IN THE GREATER LONDON AREA

Authority	*Net Expenditure on Welfare Services per 100 population*			*The % of Occupied Males, 1951 Census in Classes*			*% of Population 75 and Over 1951 Census*
	Children £	*Welfare* £	*Health* £	*I & II*	*III*	*IV & V*	
County Boroughs:							
Croydon	47	74	135	24	57	19	3·8
East Ham	50	101	156	11	58	31	3·2
West Ham	67	69	173	9	49	42	2·7
Counties:							
London	119	85	198	17	54	29	3·5
Essex	28	40	158	19	54	27	3·5
Hertfordshire	34	39	132	23	52	25	3·3
Kent	44	46	137	22	52	26	4·2
Middlesex	45	45	145	24	57	19	2·7
Surrey	35	34	127	29	52	19	4·2

Sources: Institute of Municipal Treasurers and Accountants and Society of County Treasurers, Children's Services, Welfare Services and Local Health Services, 1960-61 and Census 1951.

are best placed in the hands of the smaller authorities, a view expressed by Aneurin Bevan when introducing the National Assistance Bill in 1947—"Where the individual is concerned, where humanity and warmth is the primary consideration, there the authority which is responsible should be as near the recipient as possible."*

It may well be argued that some at least of the counties spend less on their welfare services because their populations are better off and therefore have less need of them. Some light is thrown on this in the 1951 Census (1 per cent sample), with its classification of the male population into five groups according to occupation, Classes I and II comprising the professional and managerial, Class III the clerical and artisan, and Classes IV and V the semi-skilled and unskilled. Their percentage distribution in the local authority areas in Greater London is shown in Table I. These figures give some support to the argument. On the other hand, the group of the population most subject to the need of the services is, generally, the very old, and not only the old in the lowest social classes.

It will be seen from Table I that Surrey, with the highest proportion of the wealthier classes, is also one of the two counties with the highest proportion of the old (more than half as many again as in West Ham). The county boroughs of West and East Ham, with the highest proportion in the lower social classes, are also two of the three with the lowest proportion of the very old. The county borough of Croydon ranks high both in the proportion of the wealthier and in the proportion of the aged. Both Croydon and West Ham have a proportion of children in care below the national average (see Chapter Five).

Moreover, since it is more expensive in manpower, transport and time to provide a service in areas of low population density, this increases the cost of services in counties with a large rural population as compared with the county boroughs.

If the above facts and figures suggest that the five million inhabitants of Outer London might expect to see their welfare services more fully developed locally, this does not apply to the three million in the County of London where the expendi-

* *Hansard*, November 24th, 1947, Col. 1610.

ture on health and welfare is comparable with that in the county boroughs, and that on children very much greater. Here the question is a different one, namely whether the size and scale of the organisation does not militate against the provision of a local and personal service.

The very high expenditure on the children's service must be considered against the background that the County of London has a far higher proportion of children in care than any other authority in the country—11·6 per thousand of the estimated population under 18 against an average for the rest of the country of 4·3 on 31st March, 1961. Croydon's proportion was 4·5, East Ham's 7·1, West Ham's 4·8. Moreover, four fifths of the children in the care of the County are accommodated outside the County boundary, and while there is general recognition that the boarding out of deprived children gives them the nearest substitute for a normal life, as well as being far less expensive than institutional care, the proportion boarded out for London was 23 per cent, easily the lowest in the country, as against West Ham's 40 per cent, Croydon's 56 per cent and East Ham's 67 per cent.

Again, modern opinion is insistent that where a family home is not available, institutional homes and nurseries must be as small as possible for proper child development, and in Croydon, East Ham and West Ham between them there is only one children's home accommodating more than 15—a residential nursery for 23. London has five children's homes accommodating more than 250, and 13 nurseries for more than 25, with only 11 for that number or less.

It is difficult to believe that this situation can be wholly due to special social conditions within the County, and that administrative factors do not enter in; East and West Ham are not so very different from East London. It is the aim of all Children's Departments wherever possible to re-unite child and family, but the dispersal of London children over an area up to 60 miles from Charing Cross must thwart the maintenance of family ties, with the probability of children remaining longer in care and the consequent build-up of numbers. The pressures leading to this dispersal are similar to those which have impelled Essex to provide its residential accommodation outside the Metropolitan area, and are equally likely to be more strongly resisted by a more local administration.

In the county boroughs a child care officer can and does maintain personal contact with a child she is responsible for admitting throughout its career in care. With Homes and foster parents distributed over so wide an area this is physically impossible with London children. In East Ham there are four child care officers for the 161 children in care (1—40), in London 189 for 9,000 (just over 1—50). But while in East Ham there is an administrative staff of four to support the field workers (i.e. one to one), in London the number is 283 (three to two) which is an indication of the demands on manpower for administration made by a large authority.

On the Welfare side the London County Council, although it has provided some 40 small homes since 1948, has one of the highest proportions in the country of old people accommodated in the old large institutions.* There are nine of these establishments, housing some 4,000 people, nearly all aged. The finding of alternative accommodation, or of sites for building for numbers of this order is an overwhelming proposition for any single authority, whatever its wealth and resources. Split up among 13 authorities, the problem is less daunting: a number of the county boroughs up and down the country have succeeded in closing down their large old institutions. Croydon and East Ham have none, West Ham is parting with its last at the end of this year.

Similarly it can be maintained that the problem of homelessness would not have attained the alarming proportions it has of late in London if its solution had not confronted one single authority. The 1961 Census statistics do not suggest that this problem should have been greater in London than elsewhere. In the County the number of private dwellings increased by 145,000 or 18·2 per cent between 1951 and 1961, while the number of private households *fell* by 1 per cent and the total population by 4·6 per cent. In Croydon, on the other hand,

* See Peter Townsend, "The Last Refuge", Routledge & Kegan Paul, 60/-, Table B.

both the number of private households and the population rose in these 10 years while the increase in the number of private dwellings was 4 per cent less than in London. Yet Croydon has contained its homeless problem, as have East and West Ham, where the population decreases have been greater than in London, but the increases in private dwellings smaller.

London is, of course, a Mecca to the provinces in a way the three county boroughs are not. But in so far as the problem of homelessness in London is due to an influx of provincials, the publicity given to it because of its magnitude, and the knowledge disseminated that there is just one authority available to whom application can be made for temporary accommodation, facilitate and even encourage such application.

It is the size of the problem in London and its concentration which have exacerbated the situation. Provision for the homeless has been regarded as a matter for the Welfare rather than the Housing Department and they have had to be dealt with in bulk, which has meant herding them together in large numbers in the only accommodation available to the department concerned, generally the old institutions; it is this aspect of the matter which has incurred public criticism. What can be done by a smaller authority is shown by the Borough of Walthamstow. Here the homeless are dealt with purely as a housing problem, without recourse to the temporary accommodation provided by the Essex County Council Welfare Department under the National Assistance Act. The 70 homeless families there are individually housed in substandard accommodation scattered throughout the borough.

While the foregoing facts suggest that an authority with a population of three million may be too large for the most effective operation of the welfare services, it is this very size, or rather the power and resources which accompany it, which make many reluctant to see the disappearance of the London County Council since the various committees have been willing to use these resources to experiment and pioneer in a number of directions—Welfare in provision of services for the physically handicapped and of units for the rehabilitation of substandard families and of fostering the efforts of voluntary organisations,

Children's in the provision of special homes and hostels and in the beginnings of a preventive service; Health in health centres, its central laboratory and its statistical service; all three in the provision of training facilities. But it is unreasonable to think that the new London Boroughs would permit these institutions to lapse and pessimistic to regard them as incapable of initiating further developments. As stated above, the average size of these boroughs and their resources will be greater than the majority of the 83 existing county boroughs.

There is one final reason why it appears in the interests of the welfare services that they should be operated in the London area by smaller units of government. Professor Donnison* has already pointed out the transformation that is taking place in these services; with a general movement from institutional to community care and a growing emphasis on the individual approach. The Younghusband Committee has stressed the implication of this movement—that the ever-increasing demand for social workers against a shortage of supply calls for economy in their use which requires the closest co-operation between the services. Specialisation is still needed, but much of the work is common to all the services and can be done by a general purposes social worker.

Such closely co-ordinated effort is exceedingly difficult to achieve in large authorities. Big departments tend to be self-contained and jealous to preserve their own empires. The writer once asked a welfare officer in one of the biggest provincial county boroughs what the relations of his department were with housing, and received the reply, "You wouldn't know we belonged to the same firm."

It is true that in London the Education, Health and Children's Departments have all decentralised their administration into nine co-terminous divisions. In the Welfare Department the decentralisation is to three divisions only, so that populations of about a million have one point at which to make their applications. But the lines of communications from these divisions all tend to lead to the centre on parallel lines which never

* D. V. Donnison, "Health, Welfare and Democracy in Greater London," Greater London Papers No. 5.

meet. The possibilities of cross communication are impeded because although each of the departments has its own office in the division, these are never under the same roof. In only two of the divisions are the Children's and Health Department's offices even in the same postal district, and only the welfare office is in the same postal districts as any of the others.

In Middlesex, where there is stronger feeling of local independence, the position is more favourable, for in five of the 10 local administrative areas the health, welfare and children's services are under one roof, and in others the area health offices are located in the town hall, which occurs in none of the London district offices (there are, however, three Home Help Service offices in town halls in London).

In Surrey also the Education, Health and Welfare Services are decentralised into nine co-terminous divisions, but again in only two are the Health and Welfare Offices under the same roof. The fact that Surrey's County Hall is situated at Kingston, however, well within Greater London, tends to give a greater cohesion to the services in this part of the county, and is one explanation of the reluctance of this area to be separated from Surrey. The same considerations apply to Cheshunt, which is within a few miles of the County Hall at Hertford.

The County Halls of Essex and Kent, at Chelmsford and Maidstone, respectively, are at considerable distances from their boroughs with the Metropolitan area and there is little or no co-ordinated decentralisation in either county.

Decentralisation, however essential in a large authority is a very different thing from local autonomy, and can deter initiative and lead to a feeling of frustration. Under the committee system, proposals for local improvements may have to be passed not only by the local committee, but more than one central committee (e.g. Finance and Establishment as well as the Service committee) and this can lead to protracted delays.

The ideal situation is again found in a county borough: a recently built modern extension close to the town hall in West Ham, houses three departments only, Health, Welfare and Children's, and the officers of these departments are in constant personal contact.

It is a fair conclusion to draw from all these considerations that the proposed new London boroughs not only offer the possibility of more comprehensive services outside the County boundary, and more personal service within it, but also promise a more suitable framework for future development throughout the Greater London area.

In Chapters Two to Six the present administration of these services is considered in detail as it affects the various categories of the needy enumerated on page 9, while Chapter Seven suggests how they might be reorganised in the new London boroughs.

PART TWO

CHAPTER TWO

Services for the Old

THERE is a general impression that old people in the Welfare State are fully provided for: old age and retirement pensions, the National Health Service with its geriatric units, National Assistance, meals on wheels, home helps, clubs, visiting services and old people's welfare committees must surely take care of all their needs. Yet an examination of these services in operation reveals that everywhere there is fragmentation, with gaps, inadequacies and overlapping.

At the present time, the central Government is responsible for the financial needs of the old, through the Ministry of Pensions and National Insurance by way of retirement pensions, through the National Assistance Board by way of old age pensions to the uninsured, and of allowances to those unable to manage on their pensions or without them. Central Government is also responsible in the main for the Health Service, though some responsibility also rests with the local authorities.

It is with the functions of these that this chapter is mainly concerned, though it is to be noted that some hospitals occupy themselves with the welfare of the old outside the hospital through geriatric almoners, and the National Assistance Board has a statutory duty to have regard to the welfare of all those, including the old, to whom it pays allowances. Thus there is a multiplicity of authorities dealing with the old, but there is none that has responsibility for the ascertainment of their whole needs and for the welfare of those living in their own homes.

I. WELFARE DEPARTMENTS AND THE OLD

Residental accommodation

The most conspicuous responsibility for the aged carried by

the first tier local authorities is the provision of residential accommodation "for those in need of care and attention not otherwise available to them."* The decision as to whether any individual is eligible for such accommodation rests with officers generally called welfare but sometimes, more appropriately, admitting officers, since their concern with the welfare of the old people is often limited. The administration of the residential establishments is normally the function of a welfare committee and a welfare department, this being the case with all the Metropolitan authorities except Kent, where it is that of the Health Committee and the Medical Officer of Health, supported by a Residential Services Officer.

The actual provision of residential accommodation by welfare authorities in the London area was as follows in January, 1960:

TABLE II.—ACCOMMODATION PROVIDED UNDER PART III OF THE NATIONAL ASSISTANCE ACT 1948, BY AUTHORITIES IN THE GREATER LONDON AREA

Authority	*Accommodation in large homes*	*No. of large homes*	*Accommodation in small homes*	*% of Accommodation in large homes*
County Boroughs:				
Croydon	None	None	476	Nil
East Ham	None	None	328	Nil
West Ham	150 (see text)	2	245	38 (see text)
Counties:				
London	5,064	9	1,927	72
Essex	692	5	1,296 }	35 }
Hertfordshire	332	2	701 } See text	32 } See text
Kent	226	2	2,016 }	10 }
Middlesex	444	2	2,124	18
Surrey	1,062	7	1,022	50

For the purposes of this table, larger homes are taken as former poor law institutions with accommodation for upwards of 100 persons. All the establishments are always fully occupied. Over 90 per cent of the residents are of pensionable age, so that they are virtually all homes for old people; most authorities have one or more homes for the blind, but nearly all the residents are also old; Surrey has a home for epileptics, but here too all 12 residents are aged. All authorities set aside special accommodation for the more infirm and London and Kent each have an establishment for the mildly senile.

One of the avowed purposes of the National Assistance Act,

* National Assistance Act 1948, s. 21 (1) a.

1948, was to facilitate the disappearance of the old poor law institution. In two* of the three county boroughs it has disappeared. West Ham is still responsible for the administration of two institutions, but for some years has made use of less than half the accommodation in them (150 beds out of 340) and by the end of 1962 both will be closed—one to be demolished the other handed over entirely to Essex. Thus by 1963 there will be no large poor law institutions left in the county boroughs.

The picture in the counties is very different. London has nearly three quarters of its accommodation in large old institutions, Surrey a half. Middlesex makes much the best showing with only two comparatively small institutions still in use. Kent looks better in the table, but Kent, Essex and Hertfordshire all make use of former poor law accommodation in what are now hospitals in units of less than 100 which hardly qualify for the title of "small homes". Kent has 752 such beds, Essex 330 and Hertfordshire 127.

The county boroughs do not compare so favourably in the provision of purpose built homes. West Ham has two, East Ham one and Croydon none, though Croydon has two or three nearing completion and plans others. On the other hand London has 10, Middlesex and Essex three each, Surrey and Herts two each and Kent one. Some of these new homes have been criticised on the ground that they provide a degree of luxury in which some residents are ill at ease. If adapted premises tend to be more homely, however, they rarely provide single room accommodation, which is generally much appreciated.

As well as their nature, the location of the homes is a matter of much importance to old people if they are to maintain contact with friends and relatives. All homes in Croydon and West Ham are within or just outside the borough boundaries, but only three of East Ham's eight. This is because East Ham pioneered in the provision of small homes at a time when building was impossible and houses suitable for adaptation rare in the borough.

* Croydon has a former Poor Law institution with less than 100 beds (89), which is being modernised.

The counties tend to provide their homes on a county basis, with less regard for the requirements of individual boroughs than for the ready availability of sites and premises. This applies more particularly to Essex (where the accommodation in homes for the population of 990,000 within the Metropolitan area was 770 as against 1,318 for the 760,000 in the rest of the county), and the County of London where 11 of the existing Metropolitan boroughs have no homes and another six no small homes.

This situation has been used as a strong argument against the reorganisation proposals, but it also indicates that smaller authorities pay more attention to meeting needs locally. Undoubtedly however, it will involve serious initial difficulties if and when the proposals are implemented.

Middlesex, Surrey, Hertfordshire and Kent do not present the same problem—the homes in Middlesex and the Metropolitan parts of the other counties being located more in accordance with the concentrations of population.

A further important consideration from the point of view of the old person, or his friend or relation, making application for admission to a home, is the proximity of the place at which application can be made. Conversely, a welfare officer is more likely to get the full story about a case if his cases are located within a reasonable distance of his office.

The acreages of the five authorities wholly within the Greater London Area, and the number of application points in each, is as follows:—

Authority	*Acreage*	*No. of application points*	*Average area (acres) per application point*
Croydon	12,672	4	3,224
East Ham	3,324	1	3,324
West Ham	4,689	1	4,689
Middlesex	148,684	10	14,868
London	74,684	3	24,966

Clearly the other four counties, all of which have large rural areas, are not comparable. Some comparison is however possible by taking the application points located in the Greater London area and covering boroughs and urban districts wholly within the area.

	Acreage inside Greater London covered by the application points in Column III* I	*Application points within Greater London* II	*Acreage per application point* III
Essex	78,407	2	39,203
Herts	6,934	1	6,934
Kent	54,294	7	7,756
Surrey	46,521	4	11,630

* The acreages given do not represent the acreages of these counties within the Greater London area—some application points outside the area deal with boroughs within the area, while in the case of Essex, Epping and Harlow outside the area are included in the acreage.

The welfare officers in the county boroughs are notably closer to their cases than in the counties. In fairness to Essex, which comes out worst in this comparison, it should be stated that the County Council has approved proposals to increase the application points in the Metropolitan area to four.

There is in fact a marked contrast between the various welfare authorities in the amount of welfare work done for old people by their field officers. In most cases it is limited to periodic visits to those on the waiting lists (all authorities have waiting lists for their residential accommodation). In both Hertfordshire and Kent the welfare officers in the department act also as mental welfare officers (formerly known as "duly authorised officers") and in Kent, they are in addition, responsible for the assessment and collection of charges for home helps, recuperative holidays, welfare foods and chiropody. The officer interviewed in N.W. Kent said that in these conditions their welfare work for the old was minimal—their whole energies were directed to trying to keep up with current work.

By contrast, the Welfare Department in Croydon tries to keep track of all old people in need in the borough. Two types of business reply cards are distributed wholesale—one to council departments, the National Assistance Board and other public bodies, and to voluntary organisations; the other to general practitioners. The card for general practitioners simply has space for name, address and age of the old person and particulars of assistance or service required; that for other bodies also has a space to record brief particulars of assistance given. About 100 cards a year come in, and in this way there are some 2,000 names on the register. It is part of the function of the four district welfare officers to keep touch with all the registered cases in their districts.

This scheme is of obvious value to old people in Croydon, and although there is no statutory warrant for them, similar schemes are to be found operated by a number of welfare departments up and down the country particularly in county boroughs. There is nothing similar in welfare departments elsewhere in the London area, the services for the old not being well developed in either of the other two county boroughs.

The function of welfare departments in relation to the aged is thus in fact normally limited to the provision of residential accommodation, and that of the welfare officer to deciding whether or not the applicant is suitable for such accommodation and endeavouring to secure a vacancy if he is in urgent need.

II. HEALTH DEPARTMENTS AND THE OLD

Home Helps

Although other services provided by the Health Departments, mentioned below, benefit old people, it is home nursing and home helps which affect them most profoundly. Of these, home nursing is an integral part of the health service and so does not come within the scope of this study, though it may be mentioned that in London 54 per cent and in Surrey 62 per cent of the cases nursed were 65 and over, and in Surrey they received nearly 75 per cent of the nurses' time. Also noteworthy in connection with services to the aged, is the existence in Croydon of a rehabilitation nurse, to deal with the after effects of strokes etc. The home help service on the other hand although provided under the National Health Service Act is essentially a welfare service, the home help organisers not being medically trained. In Coventry and in Northern Ireland though not anywhere in the London area, it is in fact administered by welfare departments.

An indication of the extent of the service in the London area is given by the following figures.

TABLE III.—PROVISION OF HOME HELP SERVICE BY AUTHORITIES IN THE GREATER LONDON AREA

Authority	*No. of cases helped in a year*	*Cases helped per 10,000 pop'n*	*Hours worked in a year 000's*	*Average hours per case*	*Average expenditure per 100 pop'n*	*% of long term cases*
	1	2	3	4	5	6
					£	
Croydon	2,858	115	223	77	19	70
East Ham	1,311	120	165	126	30	90
West Ham	1,161	71	196	169	24	93
London	38,031	122	5,129	135	35	85
Essex	15,688	84	2,277	147	26	76
Herts.	5,651	72	842	149	18	58
Kent	14,345	86	—	—	12	68
Middlesex	15,893	70	2,105	132	17	67
Surrey	8,902	62	1,019	114	14	54
Average		103				
Average England and Wales		68				

Sources:—Cols. 1-4 Medical Officers' of Health, Reports. 5 and 6 Institute of Municipal Treasurers and Accountants and Society of County Treasurers. Local Health Services Statistics 1960-61.

It will be noted that Croydon, East Ham and London cover more than the Greater London average number of cases, the other five counties and West Ham fewer. West Ham however, provides a high average number of hours per case, and this illustrates the dilemma of all Medical Officers of Health as to whether they shall provide an intensive or an extensive service. The proportion of long term cases, which largely comprise the old, is higher in the county boroughs and London than in the counties (except Essex).

There is still everywhere an unsatisfied demand for service, in spite of steadily increasing provision by nearly all authorities. There has also been an improvement in the kind of service since the early days, when it was virtually limited to charring. Home helps now often undertake cooking and shopping, and there are frequent tributes to their willingness to help their charges in their own time.

The attached figures from Surrey (See p. 27) indicate how the services of home helps may be allocated among the various types of recipient.

It will be seen that the "Chronic", though constituting only just over half the cases helped annually (a low proportion in

Type of case	*Number of cases helped during* 1960	*Hours of service given during* 1960	*Average total hours of service per case*	*Average No. of hours service per case per week*	*Average duration of service per case in weeks*	*Average No. of cases being helped per week*
Maternity	2,459 (27·6%)	148,721 (14·6%)	60	23·7	2·6	118 (3·7%)
Acute	1,610 (18·1%)	66,337 (6·5%)	41	7·2	5·7	174 (5·4%)
Chronic	4,725 (53·1%)	787,010 (77·2%)	167	5·2	32·0	2,859 (89·2%)
Tuberculosis	108 (1·2%)	17,243 (1·7%)	160	5·9	27·1	55 (1·7%)
County 1960	8,902 (100%)	1,019,311 (100%)	116	6·0	19·3	3,206 (100%)

Surrey as can be seen from the table above), received three quarters of the hours of help given, and comprised nearly 90 per cent of those helped in any given week. On the other hand they received a shorter time of service per week than any other category of case.

None of the foregoing figures refers to the old as such: the four categories required by Ministry of Health returns are those given above. Some Medical Officers of Health carry their analysis further however, and from their figures it appears that between 80 and 90 per cent of the "Chronic" are aged. Thus the home help service is now very largely a service for the old. In its origins, it was a Maternity and Child Welfare Service, and suggestions are still made that in times of shortage of staff, such cases get priority over the old. One authority however has a separate cadre of home helps for maternity cases.

In addition to the normal home helps, Hertfordshire and Middlesex are experimenting with a scheme, pioneered by Buckinghamshire, of "neighbourly helps". The "neighbourly help" is paired up with the person needing help, generally an aged isolate, and instead of working set hours, may get him up in the morning, help prepare a meal, and see him safely to bed at night. The "neighbourly help" has one case only, and is paid from 10/- to £2, weekly, according to the organiser's assessment of the need of the particular case.

That such schemes should be in course of development is an indication of the deficiencies of the existing home help service for the most needy type of old person.

Essex, London, Kent and Croydon provide a "night attendance service"* in connection with their home help service, with the aim of giving relief to sorely tried relatives and obviating the need for removal to hospital. In London (28 cases helped in a year) and Croydon (four) small use was made of it. In Essex also the use is limited, but it has only recently been introduced. In Kent however, 631 cases were helped in 1960 (405 for night service, 226 for evening service). The general opinion is that it is greatly appreciated when given, and would be more used if better known.

The home help service has also been developed by some authorities to serve other classes than the aged and the temporarily or permanently infirm, more particularly "problem families" (see under "Socially handicapped", page 86).

Health Visitors

Initially health visitors were concerned solely with the maternity and child welfare services, but the National Health Service Act, 1946, widened their scope, and the working party on health visitors† recommended that their main functions should be health education and social advice. As a result of this extension of functions, some Medical Officers of Health have urged that health visitors can and should serve as general purpose social workers for the old as well as for other categories.

Save in one municipal borough (Leyton, see below), there is little evidence of this attitude in the London area. There is a shortage of health visitors, authorities have difficulty in filling their establishment, and the great bulk of their work is with mothers and young children. Two additional important duties have in fact lately been imposed on them in testing infants for phenylketonuria, and very young children for hearing defects. Thus in London in 1960, out of 839,000 visits, 80,000 were classed as "miscellaneous", 14,000 as to old people (less than 2 per cent), and all the rest to expectant mothers and children under five. Few health authorities in fact indicate the number of

* This is normally provided for patients living alone, or where an aged husband or wife cannot provide the necessary assistance.

† An Enquiry into Health Visiting, H.M.S.O. 1956.

visits to the old, these being lumped in under the heading "others", generally constituting about 10 per cent of the total. The Medical Officer of Health of West Ham, however, records that "the number of elderly people needing social help and advice from the health visitor continues to increase, the health visitor visits them as often as circumstances indicate" ("other" visits in West Ham numbered 4,350 out of 35,000).

In south west Essex the borough Medical Officers of Health enjoy some measure of autonomy and three of them report the following proportions of visits specifically to the old.

Borough	*Total visits* (1)	*Visits to Old People* (2)	*% of (2) to (1)* (3)
Leyton	18,170	2,857	16
Ilford	27,364	1,073	4
Walthamstow	14,952	1,349	9

The figures in Leyton are noteworthy, and the Medical Officer there says that 1,200 old people in the borough are regularly visited by health visitors, more or less frequently as the situation demands. It will be seen however, that the average number of visits is less than three a year.

Like the home helps, health visitors also assist other categories of case in need, again more especially problem families (see page 86).

Clinics for the elderly

Clinics are run by the local health authority in Dagenham, in four areas of Middlesex and by a voluntary organisation in St. Pancras. The main purpose of these is to provide general advice on healthy living (including personal problems, diet, accidents and the cultivation of interests essential to mental health) and to identify physical or mental disabilities and then refer those attending to the appropriate agency, often the general practioner, with whom close liaison is always maintained.

In Dagenham a pamphlet on diet is distributed, and a weekly exercise class is held in connection with the clinic. In Middlesex, they have in one area arranged with the Ministry of Pensions and National Insurance that men on attaining pensionable age shall be given a pamphlet drawing their attention to the existence of the clinic.

Chiropody

Although the Cope Committee (1951) recommended that local authorities be permitted to provide chiropody, the Ministry of Health at that time ruled against the provision of this service, or its extension wherever it existed, on grounds of national economy. In Circular 11 of 1959, however, the Minister said he was prepared to approve its provision for priority classes—the old, the physically handicapped and expectant mothers.

Meanwhile, Old People's Welfare Committees throughout the London area had evolved schemes for the provision of chiropody for old people, most of which are still functioning.

The service is therefore at present in a transitional stage, but it seems probable that in course of time this will be operated as a local authority service whereupon the O.P.W. Committees now using resources for this purpose will be free to use them for something else.

There is wide variation in the extent of the provision, as may be seen by the fact that whereas Kent provided 4,659 treatments, or roughly 1,200 a month during the four months ending December 31st, 1960, London provided 186,735 or 15,000 a month, during the year ending on the same date. And whereas the proportion of men over 65 and women over 60 treated in Kent was 46, in London it was 57.

Services for the old under the Public Health Acts, etc.

It is the counties and the county boroughs who are responsible under the National Health Service Act for the provision of the personal health services, including home nursing and home helps, which particularly affect the old. But the second tier authorities—in London the Metropolitan Borough Councils, in the other counties the municipal boroughs and the district councils—also have duties which affect old people, under Section 47 of the National Assistance Act and as sanitary authorities under the Public Health Acts. Section 47 gives authority (under stringent safeguards) for the removal to suitable premises on the initiative of the medical officers of health of the aged, living in unsatisfactory conditions who are unable to look after themselves, and are not receiving proper care and attention from other persons. The Public Health Acts give sanitary authorities powers to secure the sanitation, safety and cleanliness of premises, to prevent nuisances, to cleanse verminous articles, and to cleanse at home aged persons who are unable to cleanse themselves.

To quote Miss Slack* "If prevention is the right principle, it follows that as long as the Metropolitan Borough Councils remain the authorities having responsibility for the safety, cleansing or compulsory removal of old and infirm persons, they have a very strong case indeed for providing and exercising general care for them in any way at home". "The use of L.C.C. domestic help, home nursing and health visiting services have not been found by them to be sufficient, or the division of responsibility has caused dissatisfaction and frustration which some of the boroughs have resolved by appointing and depending, wherever possible, on their own staff." As a result, by 1959 18 of the 29 Public Health Departments in the Metropolitan boroughs and the City of London had one or more persons mainly or wholly engaged on welfare work for the old. To quote again "The answer to any problem of overlapping or confusion between place or function of these workers is not to allocate a particular job to a particular person, nor to speak in general terms of co-operation, but to introduce a unified service in which one authority only is responsible for the whole field of work". This could in fact be achieved in the proposed new London boroughs.

III. VOLUNTARY ORGANISATIONS AND THE OLD

Residental establishments

While local authorities may make special provision for old people with special disabilities (the more infirm, the mildly senile, the blind) they do not provide separate establishments

* Miss K. M. Slack, "Councils, Committees and Concern for the Old"—see pp. 78-9. Codicote Press, 7/6.

for those with particular social and religious backgrounds. Voluntary organisations do so in large numbers, and all local authorities make use of them, paying agreed rates of maintenance. There are homes maintained by various Roman Catholic, Jewish, Anglican, Methodist and other religious bodies (some of which cater for other than those of their own denomination), homes for gentlefolk, homes for retired members of various professions and their widows, homes for the mildly senile. Local authorities do not generally send old people to such homes: the old person approaches or is approached by the home which then asks the authority to undertake financial responsibility, as it normally does if the applicant satisfies the statutory requirement of being in need of care and attention not otherwise available to him.

The use made of voluntary homes of all types may be seen from the following table:

TABLE IV.—USE OF VOLUNTARY HOMES BY AUTHORITIES IN THE GREATER LONDON AREA

	No. in Part III accommodation on 31.12.60	*No. in preceding column accommodated in voluntary homes*	
		No.	%
County Boroughs:			
Croydon	558	87	16
East Ham	357	27	7
West Ham	381	62	16
Counties:			
London	8,559	1,921	23
Essex	2,407	464	19
Kent	2,726	602	22
Hertfordshire	1,221	214	17
Middlesex	3,681	1,202	33
Surrey	2,707	753	28

Between 50 and 70 per cent of those for whom the local authorities undertake responsibility are aged (e.g., London 65 per cent, Kent 72 per cent and East Ham 50 per cent).

The figures indicate all local authorities resort very largely to the voluntary organisations for the provision of special types of homes. In London, Surrey and Kent, about a quarter, in Middlesex a third of all those provided with accommodation under the National Assistance Acts are in such homes. It is noteworthy that the county boroughs make less use of them than the counties.

The London County Council has power under its General Powers Acts, which other authorities have not, to make loans (as distinct from grants) to voluntary bodies for the provision of accommodation and has made extensive use of these powers, at the same time agreeing that the service of these loans shall be allowed in arriving at costs of maintenance. This has certainly enabled some homes to come into being which would otherwise not have done so owing to lack of financial resources. The abolition of the L.C.C. will mean the disappearance of this source of assistance, and it is to be hoped that some method may be found of replacing this loss.

The old in the community

The Poor Law Act, 1930, laid upon the counties and county boroughs the duty "to provide *such relief as may be necessary* (author's italics) for the lame, impotent, *old*, blind and such other persons as are poor and not able to work", i.e., they were required, in theory at least, to meet *all* the needs of the aged.

The National Assistance Act 1948, which repealed the Poor Law Act and was intended to be the final Act in replacement of it, mentions the old only three times, though Section 4 requires the National Assistance Board "to assist persons in Great Britain who are without resources, or whose resources (including benefits receivable under the National Insurance Act, 1946) must be supplemented in order to meet their requirements"—including the old. The three mentions are (1) in Section 21 of the Act which *requires* counties and county boroughs to provide "residential accommodation for persons who by reason of age are in need of care and attention not otherwise available to them", (2) in Section 31* which *permits* counties, county boroughs, county districts and Metropolitan boroughs (and the City of London) "to make contributions to the funds of any voluntary organisation whose activities consist in or include the

* Section 31 has now been amended (May 25th, 1962) to give power to local authorities themselves to provide meals and recreation for old people, in their own homes or elsewhere.

provision of recreation or meals for old people", (3) in Section 47 which empowers compulsory removal (see page 30). Thus the first tier authorities are not required to provide any service for the old except residential accommodation, and are not authorised to provide any other service (except, since May 1962, meals and recreation), but they may, in company with the second tier authorities, subsidise voluntary organisations which provide meals and recreation.

In 1950 the Ministry of Health issued to all local authorities a circular (11/50) on the welfare of old people which contained the following:

"The Act of 1948...recognised the great value to old people of mobile meals services and social and recreational clubs.... The experience gained since the Act came into force...has shown an urgent need for further services of the more personal kind which are not covered by statutory provision and which indeed can probably best be provided by voluntary workers actuated by a spirit of good neighbourliness....The essence of the matter is that all local authorities should do everything in their power to encourage further voluntary efforts to meet the needs of old people, especially those living in their own homes, and that there should be close and continuous collaboration between all the statutory and voluntary agencies working for their welfare."

Thus the position is that whereas before 1948 first tier local authorities were theoretically responsible for meeting *all* the needs of old people, since that date those needs, other than monetary assistance, residential accommodation, and the services provided under the National Health Service and Public Health Acts, have been left to the voluntary organisations, encouraged and if they so decide, financially assisted by the first and second tier local authorities.

A detailed account of how this pattern has worked out may be found in Miss K. M. Slack's "Councils, Committees and Concern for the Old" while in the Appendix (page 135) is an analysis on the same lines, though less detailed and comprehensive, of how it is working out in the 12 boroughs of Greater London in South West Essex.

A close reading of these studies is necessary for a full comprehension of the present situation, but the outstanding points are these.

Every Metropolitan Borough in the County of London, and all but one in Metropolitan Essex, has its Old People's Welfare Committee, but their scope and effectiveness vary enormously.

Nearly all are financially supported by the councils, but the amount of support varies widely in London: one council gives nothing, one less than £100, seven others less than £1,000, six from £1,000 to £1,999, eight from £2,000 to £3,999, and five over £5,000, the highest figure being £8,785. The variation is even greater in Essex, from £180 to £40,000 (see Appendix, Table B). "No taxation without representation" "He who pays the piper calls the tune"—the effect of these conceptions may or may not have been foreseen in the framing of Section 31 of the National Assistance Act and Circular 11/50, but it is clear that by reason of the contributions made from the rates and the part played in them by borough councillors and officials in some boroughs (see Appendix Tables B and C), the O.P.W. Committees are virtually departments of the council. Whether this is desirable or not depends on the point of view.

Because of the stress laid on meals and recreation in the Act, it is these services which have been most completely developed. Practically every borough has its cheap meals, all have a number of clubs, and most provide cheap holidays. But generally speaking, apart from those receiving meals on wheels, it is the active and mobile and socially minded who principally benefit from these. Those most in need, the housebound and the isolated, cannot profit from them.*

Mobile meals are undoubtedly one of the most valuable and appreciated services. It is on the whole those who really need

* Cf. the Herbert Report (Royal Commission on Local Government in Greater London, 1960), p. 161, which observes "a tendency for too much of the limited funds available to be spent on providing clubs, entertainments and subsidised seaside holidays for old people many of whom are still hale and hearty, and a proportion of whom are not in financial straits, and not enough done to identify and help the frail or solitary who need a regular visiting service, meals on wheels, or an all-day club providing warmth, food and interest".

them that get them. But in London a five day a week service which is generally what is needed is the exception, and in Essex it is provided in only one borough. This accords with the findings of a survey* of the service over the country as a whole, which concluded that only a quarter of the total need was being met. The main obstacle to development is the lack of sufficient voluntary personnel and transport and some boroughs to meet this situation now provide paid drivers, volunteers undertaking the delivery of the meals.

The other services of greatest value to the most needy is visiting, by which is meant not social visiting, though this has its place, but the visiting which is made to deal with a known or suspected need. Pages 146-149 of the appendix indicate what this sort of visitation can achieve and also the need for its performance by a whole-time skilled practitioner. There are only two such whole time trained visitors in Essex, both borough council employees, but both working closely with the O.P.W. Committee (of one of which the Medical Officer of Health was chairman, and the other the Clerk of the Council). Full time workers are found much more frequently in the London boroughs (Councils, Committees and Concern for the Old, Table 4, pages 142/3), but their conditions of employment vary widely, some being directly employed by the O.P.W. Committees, and others officers of various departments of the councils.

Generally speaking, it is in those boroughs where the council is most actively interested in the welfare of old people that the old people most in need get the best service. In these circumstances it would seem that the time has now come to give local authorities a definite responsibility for their well-being. This could most effectively be done, not by merely empowering them to provide meals, etc. (National Assistance 1948 Amendment Act 1962), but by amending subsections 1 and 4 of Section 29 of the National Assistance Act 1948 to read (alterations *in italic* and omission bracketed):

S. 29 (1) "A local authority shall [have power to] make

* Amelia Harris. "Meals on Wheels for Old People," National Corporation for the Care of Old People, 7/6.

arrangements for promoting the welfare of persons to whom this section applies, that is to say persons who are blind, deaf or dumb, and persons who are substantially and permanently handicapped *physically, mentally or socially,* by *old age,* illness, injury, or congenital deformities, or such other disabilities as may be prescribed by the Minister". S. 29 (4) (f) "for providing such persons with recreational facilities *and meals,* in their own homes or elsewhere".

This would have the effect of confining the responsibility of the authorities to handicapped old people—they would have no responsibility for the able bodied woman of 60 or the man of 65,* the usual definition of old people being those of pensionable age.

The work of the voluntary organisations could continue exactly as at present. They could if so desired act as agents of the authority under section 30 of the act, could continue to receive grants under section 31 in providing meals and recreation for the old not handicapped, and could continue to co-operate with the authority in aiding the handicapped old.

Presumably the new London boroughs will be constituted welfare authorities under the reorganisation. The counties already have powers, under the Local Government Act 1958, to delegate their section 29 functions to district councils with a population of not less than 60,000 outside the Royal Commission review area. It is to be hoped that in time, if the Act were amended as suggested, powers to care for the old might be extended to all housing authorities. This would be a reasonable step to take, since the old needing care greatly outnumber the physically handicapped, and the needs of the old are likely to be best served when their welfare and housing are the responsibility of one authority (see Chapter Six).

* To stop the issue of meal vouchers to persons of pensionable age simply in order to supplement their pensions (see Appendix, page 141), an amendment might also be made, to s. 29 (6) (a), to read:

"Nothing in the foregoing provisions of this section shall authorise or require

(a) the payment of money, *or vouchers in lieu of money,* to persons to whom this Section applies . . .".

CHAPTER THREE

Services for the Physically Handicapped

SECTION 29 (1) of the National Assistance Act 1948, gave the first tier local authorities permissive powers to "make arrangements for promoting the welfare of persons who are blind, deaf or dumb, and other persons who are permanently handicapped" but section 29 (2) also authorised the Minister of Health to require local authorities to undertake these services as a duty. And in fact the Ministry circular 87 of 1948 at once made the provision of services for the blind a duty, as did circular 15 of 1960 the provision of services for the deaf or dumb and the permanently handicapped. The Act requires that the services should be carried out in accordance with a scheme, and a model scheme was issued in circular 32 of 1951. The dates when the schemes of the authorities in the London area were approved, the numbers on their registers, and their net expenditure per 1,000 population on the blind and physically handicapped, are given in Table V.

The noteworthy features of these figures are the delay of some authorities (notably Hertfordshire and Surrey) in submitting their schemes; the similarity, in relation to population, of the numbers of registered blind in the counties, and their higher proportion in the county boroughs and London; the wide divergencies in the numbers of the other handicapped with the county boroughs again showing the highest proportion registered, and the higher expenditure on these services of the county boroughs (except West Ham on the physically handicapped). These points are further enlarged on below. In all authorities but one, the Welfare Committee is responsible for policy in regard to the handicapped and in all but two the Chief Welfare Officer for administration. In Kent the Health

TABLE V.—PROVISION OF SERVICES FOR THE PHYSICALLY HANDICAPPED BY AUTHORITIES IN THE GREATER LONDON AREA

Authority	*Date of Approval of scheme for Physically Handicapped*	*Numbers on Register,* 31.3.61*						*Net Expenditure per 1,000 population on:*	
		Blind and partially sighted		*Physically Handicapped*		*Deaf and Dumb*	*Hard of Hearing*	*Blind*	*Physically Handicapped*
		Total	Per 1,000 population	Total	Per 1,000 population			£	£
Croydon	6.12.52	861	3·5	551	2·2	13	10	65	36
East Ham	12.3.52	380	3·4	392	3·6	39	19	101	50
West Ham	31.10.55	459	2·8	344	2·1	199	6	98	5
London	9. 4.52	10,023	3·1	6,027	1·9	123	318	56	31
Essex	29. 7.53	4,459	2·4	3,280	1·8	754	317	30	19
Hertfordshire	7. 1.57	1,642	2·2	358	0·4	342	318	24	7
Kent	12.11.55	4,087	2·2	587	0·3	717	325	21	6
Middlesex	14. 7.52	4,961	2·4	3,858	1·3	1,532	682	37	22
Surrey	10.12.58	3,150	2·2	2,600	1·8	—	—	22	8

* The National Assistance Act requires authorities to keep a register of the number of cases assisted. The figures given are thus not the numbers of handicapped in the authority's area, but the numbers assisted by the authority.

Committee and Medical Officer of Health are responsible, in Surrey the Welfare Committee and the Medical Officer of Health.

I. THE BLIND

Of all the handicapped groups in the population, it is the blind who everywhere receive the most comprehensive and effective service, from both statutory and voluntary bodies. Their disability is obvious and dramatic, and readily enlists the sympathy of the public, which has meant not only that the voluntary societies are richly supported, but also that legislative action was taken early on their behalf. The first blind Persons Act was passed in 1920, and this besides giving the blind pensions at 50, took them out of the ambit of the Poor Law, then administered by the boards of guardians, and placed their welfare in the hands of the first tier local authorities. An Act in 1938 lowered their pensionable age to 40, and prohibited the giving of outdoor relief to the blind under the Poor Law (in the interval since 1920 the first tier authorities had become the Poor Law Authorities). The National Assistance Act, 1948, repealed Sections 2-4 of the Blind Persons Act, 1938, and allotted their welfare to the local authorities, assimilating them to the other handicapped in Section 29, but did not repeal the early granting of old age pensions (Section 1 of the 1938 Act). The Act, in Section 5 (2), further required the National Assistance Board to make special (i.e., specially generous) provisions for the blind, who also received a special concession in Section 7 (6) (g).

The blind have therefore been particularly favoured, not least by the fact that the local authorities have had 40 years of experience of operating welfare services for them.

Initially it was a general practice among county councils and county boroughs to employ voluntary societies to carry out their duties to the blind, but now most of them perform them through their own officers, Hertfordshire being the only authority in the London area to employ an agent (the Hertfordshire Society for the Blind) to which it gives substantial financial support. All authorities however, make extensive use of the facilities offered by the voluntary societies (see below).

A high proportion of the blind are old people. In London, 67 per cent of the 1,168 newly registered as blind during 1960 had gone blind after the age of 70. In Kent, of 3,347 blind on the register in 1960, 2,600 (77 per cent) were over 60 and 2,057 (62 per cent) were over 70. In total, because women are longer lived, blind women outnumber blind men by nearly two to one, but in the age groups under 60 the men outnumber the women. These figures refer to the blind only (those so blind as to be incapable of any work for which sight is required) and not the partially sighted (those permanently and substantially handicapped by defective vision) both of which categories are included in the numbers registered given in Table VI. The usual ratio of partially sighted to blind on the register is from a quarter to a sixth.

The backbone of the blind welfare service is formed by the home teachers of the blind, whose main functions are to visit the blind and partially sighted in their own homes, teach them where practicable to read embossed literature, teach them pastime occupations, help them overcome the effects of their disabilities and assist in promoting their welfare generally.

All authorities have appointed teams of home teachers but their numbers in relation to the numbers of registered blind vary considerably, which is one factor in their differing expenditures.

TABLE VI.—Home teachers of the Blind with Authorities in the Greater London Area

Authority	*No. of registered blind and partially sighted*	*No. of home teachers*	*No. of reg. cases per teacher*
County Boroughs:			
Croydon	861	6	143
East Ham	380	3	126
West Ham	459	3	153
Counties:			
London	10,023	49	204
Essex	4,459	24	186
Hertfordshire	1,642	9	162
Kent	4,087	16	255
Middlesex	4,961	30	165
Surrey	3,150	16	197

It is clear from the figures given regarding ages that after everything possible has been done to teach the newly blinded how to overcome the effects of their disability, the home teachers cannot do much more with the majority of their cases than visit them socially, which is an extravagant use of their skills.

Much more can be done, and is done, both by the home teachers and other agencies, for the younger blind and for the newly blind. In addition to visits, three month non-residential rehabilitation courses are held for the newly blind in London, and the Royal National Institute for the Blind provides residential courses for the same purpose, which all the authorities make use of.

Wherever possible, those capable of work are found employment in open industry, in sheltered workshops or at home. Training courses are arranged where necessary, through the Ministry of Labour.

The counties of London (3) and Essex (2) have their own placement officers (one in each case being himself blind). Some 700 blind are working in open industry in London. Other authorities rely on the placement service provided by the R.N.I.B.

London and West Ham each has its own workshop for the blind, London's having been taken over from a voluntary society. West Ham reserves a certain number of places for other authorities. All authorities however, including London make use of the workshops of the various voluntary organisations, some of which provide residential accommodation for both men and women in connection with them.

Other blind workers, capable of earning a minimum wage, are employed on Home Workers schemes, also organised by voluntary bodies. Both these and workers in the blind workshops have their wages augmented by the local authorities, to a certain sum which varies from time to time.

The voluntary societies also maintain a number of homes extensively used by the local authorities and all authorities also support the National Library for the Blind, which circulates embossed literature.

The social needs of the blind are taken care of both by the local authorities and by the voluntary bodies. Thus the L.C.C. has a senior pastimes officer and runs some 50 clubs and classes (e.g., in musical appreciation) for the blind in 18 different premises, while in all areas there are local associations for the blind which cater more specifically for their social needs.

The quality and quantity of service for the blind is impressive, both by reason of comprehensiveness in all areas, and also because of the close co-operation both between the local authorities and the voluntary societies, and among the societies themselves, where each now has its pretty clearly defined sphere of operation. Less stress is generally laid on the services provided for the partially sighted, and the Ministry of Health report for 1960 suggests that their needs are on the whole, less fully recognised and less provided for, adding that much is still needed to decide how their need can best be met, and that it is not always appreciated that the most useful service which can be given for some of the members of the group may be one designed to provide—with suitable safeguards—that the fullest use is made of residual vision. There is now a move on foot to assimilate the partially sighted, other than those likely to go blind within four years, to the classes of physically handicapped—to take them from the sphere of the blind home teachers to that of visitors to the physically handicapped.

II. THE DEAF AND HARD OF HEARING

The needs of the deaf and hard of hearing are very different from those of other types of handicapped. On the one hand employment scarcely presents a problem; it having been estimated that about 90 per cent of the employable deaf are in work. On the other hand, because they lack the normal means of communication, they can only enter fully into social activities arranged by and among themselves.

It is for this reason that all authorities employ the Associations for the Deaf and Dumb and the Leagues of Hard of Hearing as their agents rather than undertake the work themselves. Thus London, Essex, Middlesex, Surrey, West Ham,

Croydon and Kent employ the Royal Association in aid of the Deaf and Dumb, Kent the Canterbury Diocesan Association and Hertfordshire the St. Albans Diocesan Association for the Deaf. London supports the London League for the Hard of Hearing; Surrey, Middlesex, Croydon and Essex the Surrey and Middlesex League; Essex also the Essex League; Herts. the Herts. League. East Ham is unique in having its own Club for the Deaf, of which the chairman and vice-chairman of the Welfare Services Committee are president and vice-president respectively, and the chief officer and his deputy chairman and vice chairman. The club meets twice weekly in the Council's occupational centre (see p. 48) and although mainly self-supporting, receives council help with administration.

The disparities in the figures of the number of deaf on the registers in the table on p. 39 seems largely due to whether or not they include those helped indirectly through the Associations. Thus in Middlesex on 31st December, 1961 only 254 deaf and hard of hearing were registered with the Council direct, but in addition 1,135 Middlesex cases were known to be registered with the R.A.D.D. and 669 with the League for the Hard of Hearing.

In addition to the local associations for the deaf and hard of hearing, the National Institute of the Deaf and the British Association for the Hard of Hearing advance the interests of these groups on a national scale, and the N.I.D. also provides certain residential establishments.

III. CRIPPLES AND OTHERS ("GENERAL" CLASSES)

Both the blind and the deaf form comparatively homogeneous groups, and both have had services available to them for a long time, the blind from both statutory and voluntary, the deaf from voluntary sources (the R.A.D.D. was founded in 1840).

The other physically handicapped are not a homogeneous group. The Ministry of Health has prescribed a classification of disabilities and the following tables show the numbers under each heading as recently shown in the returns of a county and a county borough.

TABLE VII.—CLASSIFICATION OF DISABILITIES OF THE PHYSICALLY HANDICAPPED IN MIDDLESEX AND EAST HAM

Classification	*Middlesex* *Nos.*	*Middlesex* %	*East Ham* *Nos.*	*East Ham* %
1. Amputation	229	4	5	1
2. Arthritis and Rheumatism	1,221	23	55	15
3. Congenital malformations and deformities	114	2	3	1
4. Diseases of the digestive and genito-urinary systems, of the heart and circulatory system, of the respiratory system (other than tuberculosis)	866	16	41	11
5. Injuries of the head, face, neck, thorax, abdomen, pelvis or trunk; injuries or diseases (other than T.B.) of the upper and lower limbs and of the spine	475	8	71	19
6. Organic nervous diseases—epilepsy, disseminated sclerosis, polio, etc.	1,445	27	94	26
7. Neuroses, psychoses and other nervous and mental disorders	147	3	66	18
8. Tuberculosis (non-respiratory)	69	1	2	1
9. Diseases and injuries not specified above	850	16	28	8
	5,416*	100	365	100

* The Middlesex total includes cases of multiple handicap. The number of individuals concerned was 4,304.

The Middlesex analysis goes on to analyse further those in categories 8 and 9 with the following results:

8. Epilepsy	259	Poliomyelitis	342	Parkinson's Disease	106
Sciatica	1	Cerebral Palsy	264	Syringomyelia	14
Multiple Sclerosis	442	Fredrick's ataxia	17	Cancer	29
9. Cerebral Haemorrhage	371	Muscular dystrophy or atrophy	69	Nephritis	7
Encephalitis	20	Tuberculosis (respiratory)	52	Sundry	302

These details indicate the almost bewildering variety of disabilities for which the service has to cater. They also help to account for the multiplication of voluntary organisations in this field, since many of them cater for the sufferers from one particular disability.

There are however, other factors which tend to simplify the situation. In the first place the employment of the physically handicapped in open industry and in sheltered workshops is, under the Disabled Persons (Employment) Act, 1944, the concern of the Ministry of Labour in association with Remploy, the government sponsored organisation for the provision of sheltered employment. A recent survey by the Ministry of Labour indicates that 96 per cent of all handicapped school leavers are found employment. As will be seen below, there is some overlap with local authorities, who also provide sheltered workshops, but in general the responsibilities of the latter, in the words of the Piercy Committee,* are to meet the social and occupational needs of those disabled who do not come within the employment field, and simultaneously to cater as far as possible for the social needs of the disabled in the employment field.

Secondly, as is emphasised elsewhere, although particular disabilities may involve particular needs, the basic needs of all handicapped persons are the same, whatever their disability—adequate and suitable food, clothing and housing, companionship, recreational and cultural facilities, the self respect which is in essence a consciousness of the ability to give as well as get.

The total number of blind in the London area probably approximates fairly closely to the number registered by the local authorities. The number of the "general class" of the handicapped (i.e., those other than the blind, deaf or dumb) certainly does not—those registered consist simply of those whose needs have come to light and have been met in some way. It has been suggested† that the total number of the permanently and substantially physically handicapped in the County of London may exceed 150,000, many or most of whom may not be in need of advice or assistance. But the scope for development may be gauged from the fact that whereas in 1955 there were only about 1,500 of this class on the London register, by 1961 there were over 6,000, and today there are over 7,000, the number increasing by something approaching 1,000 annually. It will be seen from Table V that in most authority areas including London, the number of registered handicapped is

* Report of the Committee of Enquiry on the Rehabilitation Training and Resettlement of Disabled Persons, 1956, Cmd 9883.

† Welfare in London. London County Council, 1955.

about 60 per cent to 70 per cent of that of the registered blind. In East Ham it exceeds this number, in Hertfordshire it is about 20 per cent and in Kent less than 15 per cent of the number. It is clear therefore, that there is a very considerable amount of undiscovered need, especially in the two last named counties, both comparatively late in producing their schemes.

Of the services which can be rendered to the disabled, the first is the ascertainment of the disability followed by instruction in the means of overcoming the handicaps imposed by it, and assistance in obtaining the necessary aids. This may involve the supply of specialised equipment or "gadgets", or the adaptation of living quarters by the widening of doors, the provision of ramps and handrails and so on. All this implies a visitation service, with visitors skilled in the knowledge of needs and possibilities.

Second, there is the provision of occupation and interest to those cut off from the normal stream of life, and this can be done in the disabled person's own home by a visitor, or in special centres or workshops and in clubs, in all of which there at once arises the question of transport and specialised transport at that. An indication here follows of how, and how far, these services are provided by the authorities in the London area.

In *Croydon* the services are under the direction of a handicapped persons welfare officer, who is a trained occupational therapist, with her headquarters at the Council's own craftwork centre. Initial visits are made by one of the four district welfare officers and the handicapped persons welfare officer then follows up the visit where necessary. The number of the registered handicapped, 551 on 31st March, 1961, and 590 three months later, is said to be growing rapidly.

The Council's own craftwork centre, and that run by the Croydon Council of Social Service, accommodate 25 and 18 workers respectively. Some attend daily, others two or three days a week and 140 are in attendance in all. Each centre has two instructors, and one an occupational therapist also. Thirty-seven different crafts are taught and out work is taken in from local firms in substantial quantities, a special and successful effort having been made to develop this outlet.

Workers are brought in by a special bus, and a midday meal is served at 1/-.

A combined workshop for the physically and mentally handicapped, to accommodate 100 of each at a cost of £90,000 is in preparation, sponsored jointly by the Health and Welfare Committees, but to be administered by the Welfare Department.

The Croydon Council of Social Service has an active handicapped section, working in close association with the Welfare Department, with whose approval it has provided the (purpose built) second craftwork centre. It also has an occupational therapist who visits the handicapped in their own homes, a service not provided by the Welfare Department. Two clubs are run for the physically handicapped, one for those up to 40, the other for those over 30, the overlap being intentional. Each meets once a week. The social needs of the handicapped apart from this are met by the other voluntary organisations dealing with them generally (Croydon Association for the Physically Handicapped) or with specific handicaps (local branches of the Infantile Paralysis Fellowship, Multiple Sclerosis Society, Muscular Dystrophy Group, National Spastics Society). All these are in receipt of grants from the county borough council.

The handicapped section of *East Ham* Welfare Department consists of four administrative officers, three home teachers of the blind, and one woman welfare worker. The blind visitors deal with the blind only, but any of the others may visit and otherwise deal with the handicapped. The work centres round the occupational centre, attended at the date of visit by 68 blind and 63 other handicapped, the average daily attendance being about 50 handicapped and 40 blind in two sessions. A coach is used for transport and a midday meal provided at 8d. There are two instructors for the handicapped, and the products are of the traditional kind—baskets, trays, stools, knitting and a certain amount of elementary carpentry. Fortnightly entertainment is given to the handicapped in the centre, there being a well equipped stage.

The Department maintains an "East Ham Handicapped

Persons' Fund", run voluntarily by officers of the Welfare Department and raising about £800 a year by subscriptions and special efforts, its object being to provide recreational facilities and financial assistance to the blind and handicapped. In 1961 500 handicapped (including escorts) were taken in 11 coaches for a day's outing to the seaside by means of this fund.

In *West Ham* the Council has an agency arrangement with the "West Ham County Borough Handicapped Persons Voluntary Welfare Committee." This Committee runs an occupation centre two days a week with an attendance of 40-50 in a hall maintained by a local charity, with the Welfare Department occupational therapist for old people's homes in attendance. A special coach for transport has been presented to the centre and is operated voluntarily by local bus drivers.

The Committee has no staff of its own, but the Deputy Chief Welfare Officer is Secretary, and much of the work is done in the office of the Welfare Department. A limited amount of visitation of the handicapped in their own homes is done by the Committee in theory, but not much in practice. The Chief Welfare Officer is anxious to get more of this done by his own officers, but has not the staff as yet. The last column of the Table on p. 39 tells its own story here.

In the *County of London* the services for the physically handicapped which have been considerably expanded of late, are carried out in the General Division of the Welfare Department, and the field work is organised in three districts, each covering a population of about a million, centred on offices in Westminster, Islington and Bermondsey. Each district has an administrative officer in charge, each with admitting officers under him, and two teams of social workers, each team taking half the district, the total number working for the physically handicapped (including the partially sighted) in all the teams being 62. Visitation and attention to the individual needs of the handicapped is undertaken by these teams of social workers. Each of the three districts has a rehabilitation centre (one being purpose built at a cost of £130,000), open five days a week, associated with the district office, and there are a further nine such centres in various parts of the county. The attendances average 40-45 at the three district centres and about 24 at the others. Not all attend every day, and the total role is about 600. In the centres the disabled are given instruction into how to overcome the effects of their handicaps and craft teaching and occupation is provided. A midday meal is cooked and served on the premises at 1/-. Transport is effected by 14 specially constructed coaches, which incorporate a lifting platform, plus three utilibuses.

The department also maintains a workshop employing 45 disabled persons unacceptable in open employment or to Remploy. Joinery is the staple occupation, and it is claimed that the value of the output per head per annum at £450 is equivalent to that achieved in the Remploy factories. The social needs of the handicapped are taken care of by the voluntary organisations, but accommodation and transport are placed at their disposal, and financial assistance is given for holidays.

Initially, *Essex* employed the Essex Association for the Welfare of the Physically Handicapped as its agent for the performance of all its services for the physically handicapped, having in fact, in collaboration with the Central Council for the Care of Cripples, taken a leading part in the formation of this Association.

In 1958 however, the Council decided that while continuing to support the work of voluntary organisations in this field, the main work of providing the services would be developed through its own staff. At the end of 1961, 11 home visitors to the handicapped had been appointed and an increase is contemplated. Two daily handicraft centres were in operation, both in the review area, and others were in preparation. The Council however, continues to give substantial financial support (over £8,000 in 1960/61) to the Essex Association for the Welfare of the Physically Handicapped, which is very active in promoting local associations and clubs for the handicapped, the sale of work produced by them, and the provision of holidays. Clubs exist in all the boroughs in Metropolitan Essex, transport to them being arranged locally where necessary; some 4,500 articles were sold and £1,300 paid to the makers and 700 disabled were sent on holiday in 1961.

The service in *Hertfordshire* is run through the Hertfordshire Association for the Welfare of the Physically Handicapped, of which the Council pays the administrative expenses. The Association is administered through seven self-supporting district committees, one of which contains Barnet and East Barnet.

The Council also employs one craft instructor and home work organiser, seconded to the Association; has given the Association one specially adapted vehicle for transport and a stock of aids; financed some of the 78 holidays provided in 1960/61, and is making a grant (as is the Ministry of Labour) towards the capital and running costs of a sheltered workshop for 30 men and women in preparation in Watford. There are three work centres in operation, none in the review area, and a number of clubs, including one each in the Barnets, but these meet once a month only.

In *Kent*, the 21 district officers of the Health Department (each of whom has an assistant) have been designated as welfare officers for the physically handicapped. They are also responsible for admissions to Part III accommodation, act as mental welfare officers (formerly duly authorised officers) including after care, except for women and boys under 16, and are responsible for the assessment and recovery of cost of home helps, welfare foods and chiropody.

One craft instructor was appointed in 1959, and the Council in 1960 approved an increase of establishment to seven, but it had only been possible to fill one vacancy up to March 1961. Holidays for eight handicapped persons were provided during the year, and financial assistance given to voluntary organisations in respect of 15 clubs.

In 1961, the Kent Association for the Welfare of the Physically Handicapped was revived with the assistance of the Central Council for the Care of Cripples.

In *Middlesex* the services for the disabled are operated direct by the Council through the Welfare Department. The work is decentralised into 10 areas, and in each there are two or more welfare visitors to the handicapped (22 in all) and one home work organiser (10).

There are eight occupation centres in operation, with an average attendance of about 25 and in these considerable success has been achieved in getting out-work with the result that over the past year some £20,000 was paid over to those attending the centres. Two special coaches are provided for transport.

The social needs of the handicapped are taken care of both by the Council and by voluntary organisations. The Council through the Education Department maintains 10 choral classes (a choral festival is held annually, attended by about 200) and eight dancing classes for the handicapped. The Red Cross runs 22 handicraft classes and receives a grant for transport. Other voluntary organisations running clubs can receive 75 per cent of the cost of hire of halls and of transport costs, and the special coaches are lent on occasion. Holidays are provided for 300-400 annually at a minimum charge to the handicapped of £1 (expenditure up to £4,000 is approved of which about half is raised from charitable sources). Those making their own holiday arrangements can receive transport to a main line station.

Surrey did not produce its scheme for the welfare of the general classes of the handicapped until 1958, and has implemented it by entering into an agency agreement with the Voluntary Association for the Surrey Disabled, which exercises the mandatory clauses of the scheme and also provides handicraft classes, aids for the handicapped, holidays, transport, etc.

Within the Council, the Medical Officer of Health is responsible for administration to the Welfare Committee, and the occupational therapists, originally appointed for the tuberculous, now serve the other handicapped also, who take up an increasing proportion of their time as their number increases and the number of tuberculous diminishes.

The team consists of occupation therapists, a technical instructor, a supplies and marketing officer and a clerk-storekeeper. One hundred and twenty-three handicapped other than tuberculous were on their registers in 1960. The Council also gives direct assistance to some handicapped for adaptations etc., the amount so spent in 1960/61 being £671, while the contributions to the cost of administration to the Voluntary Association for the Surrey Disabled was £1,166, and to other voluntary associations £371.

The details above show there is a wide variation in the quality and extent of service given to the general classes of handicapped in the London area, and that a slender provision is associated with voluntary organisations acting as agents. Here the experience of Essex is significant. The Essex Voluntary Association continues to perform valuable functions with the substantial support of the Council, but it has been mutually agreed that with very few exceptions, local associations cannot maintain an effective visiting service, and, as with old people, it is visitation by skilled workers which is the key to success.

Voluntary organisations

Nearly all special accommodation for the handicapped is provided by voluntary bodies. The only provision made by the local authorities, apart from Surrey's home for elderly epileptics, is a small unit for younger handicapped men in one of the large homes in London.

The use made by local authorities of voluntary homes in general is indicated on p. 32. The use of such homes for the physically handicapped by the London County Council as on 31st March, 1962, is shown in Table VIII. By this time the number in all homes had increased to 2,184, as compared with the 1,921 shown in Table IV.

TABLE VIII.—County of London—Number in Voluntary Homes
31st March, 1962

	Men	*Women*	*Total*	
Aged	367	1,041	1,408	
Blind	44	110	154	
Cripples and Incurables	184	156	340	
Deaf and Dumb	14	19	33	
Epileptics	29	36	65	
Unmarried mothers	—	177	177	(plus 94 babies)
Others	4	3	7	
Total	642	1,542	2,184	,,

The 340 "cripples and incurables" were accommodated in 49

different homes, including those with productive workshops for younger people, those for particular types of disability (e.g., spastics) and those, such as the Cheshire Homes, for the seriously disabled and semi bed-ridden.

The National Association for the Paralysed, just before its dissolution in 1961, produced a booklet entitled "Permanent Accommodation for the Physically Handicapped" embodying the results of a questionnaire to local authorities, 80 per cent of whom replied. Eleven of them said they had special homes for this class, 116 said that they had not. One hundred and twenty-six made use of voluntary homes, five did not and 74 said they would make more use of them if they could get the beds, against 37 who said they would not. One hundred and four agreed there was a need for a classified list of such homes. The N.A.P. summed up by saying it hoped the pamphlet would "cause some steps to be taken to permit this intolerable situation from deteriorating further. Sufficient to say here that waiting lists are long, that this enquiry has confirmed yet again that Part III accommodation is inadequate, and that there is insufficient interchange between hospital boards and local authorities which results in beds being blocked by unsuitable people in both types of accommodation."

In this connection it may be mentioned that the City Parochial Foundation has undertaken to build in Essex, in association with the County Council, a model home for cripples which might serve as a prototype for other local authorities wishing to make their own provision.

As to voluntary services for the handicapped in the community, this has recently been the subject of an enquiry by Dr. J. H. Nicholson sponsored by the National Council of Social Service and the Central Council for the Care of Cripples.* The main conclusion which emerged from this study was that voluntary bodies needed to work out practical schemes of co-operation among themselves, and also give greater consideration to their relation, "individually and perhaps corporately," with the public authorities, central and local. The warden of

* J. H. Nicholson, Help for the Handicapped, National Council of Social Service, 1958,

Queen Elizabeth's College for the Disabled has expressed similar views.* "Although government departments are often criticised for lack of co-ordination, it can be suggested that there is even less co-ordination among voluntary bodies . . .

If it is difficult sometimes for a voluntary body to know which department to approach, it must be at least equally difficult for a department to know which voluntary body to consult . . . Both the Central Council for the Care of Cripples and the British Council for Rehabilitation arrange courses on various subjects, but it can be argued that these two bodies provide duplication of services and would do well to amalgamate."

And the final report of the National Association for the Paralysed commented as follows: "The most significant aspect of this classification" (that of new cases added to N.A.P's register between 1st April, 1958 and 1st April, 1961) "is that too large a proportion of the problems referred to N.A.P. have been on behalf of people with diseases or disabilities for which there are specialist organisations. Presumably this was because these organisations did not provide the particular service that was required or the information that was sought. What then is the function of these specialist national organisations? Are they really doing what is essential? Do they know what each is in fact doing or planning to do?

"Until a satisfactory answer is given to these questions, the cry will still go up that there are too many voluntary organisations; that there is no co-ordination between them; that individual cases are often dealt with simultaneously by two or more of them, unknown to each other; and that the number of appeals made by all these bodies is unending."

Ideas of a rationalisation of the services for the handicapped and their better co-ordination have been launched. There has been some response to them. The Standing Conference on Voluntary Organisations set up a working party on the Nicholson Report, which has presented three progress reports and recommended a national conference on local co-operation. The Ministry of Health has produced a circular (7/62) urging

* A. E. R. Bruce, O.B.E., Royal Society of Health Journal, Vol. 79, No. 4. July-August, 1959.

co-operation. Practical results may follow. But it does seem that there is too much talk about co-operation, and too little thought about the relative functions of voluntary and statutory bodies in this field, on the lines of the suggestions by Miss Slack in regard to old people.*

What voluntary associations for the disabled run on a local basis cannot do, because they have not the financial resources and cannot offer career prospects, is maintain a staff of visitors of the quality and quantity required for adequate ascertainment and assessment of need, nor can they provide more than a minimum of special transport. Where such work is effectively done in the Greater London area, it is done by Council staff.

What the local and specialist voluntary organisations can do is assist in all manner of ways to meet the needs when ascertained and assessed. But to do this effectively, each organisation must know what the others are doing, and it would seem natural since it has a duty to the disabled under the Act, and is responsible for keeping the register, that the Welfare Department of the local authority should take the lead in seeing this is achieved.

An example of this pattern of co-operation is to be seen in the County of London, where a Central Consultative Committee on the Care of Physically Handicapped was established by the Welfare Department in the early days of the operation of its scheme. This committee, of which the chief officer of the Welfare Department is Chairman, comprises nine representatives of voluntary organisations, two of the Welfare Department (including the chairman), one each of the Education, Housing and Public Health Departments, with observers from the National Assistance Board, the Ministry of Labour, and the Ministry of Health. It meets about twice a year, and reports annually to a conference (chaired by the Chairman of the Welfare Committee) with a much wider representation. The clerking of both committee and conference is done by the Welfare Department. The committee considers a wide range of subjects affecting the disabled and recommends suitable action by the appropriate body, official or voluntary. During the past year for example, a register of all clubs for the handicapped in

* Op. cit. p. 111.

the county has been compiled, information regarding ground level public conveniences circulated, the need for the provision of a holiday home and the means of financing it examined, a July conference of club leaders arranged, and a scheme of car parking badges for cripples agreed with the Metropolitan Police.

The disappearance of this committee with the disappearance of the London County Council would be a severe loss, and suggestions are made in Chapter Seven as to how this might be obviated.

The foregoing pages, and more especially Table V, show that ascertainment of the blind and the physically handicapped appears to be most complete in the county boroughs in respect of the blind though it is also high in the County of London. East Ham has the highest proportion registered of both blind and handicapped and one of the most fully integrated services for the handicapped is found in Croydon. It is also significant that in the County of London it has been found desirable to organise these services in six districts instead of the three into which the County is divided for other Welfare services.

All this would seem to indicate that these services are better organised on a more local basis, such as would be afforded by the new London Boroughs.

CHAPTER FOUR

Services for the Mentally Handicapped

I. THE MENTALLY ILL

DURING the past generation a revolution has taken place in the attitude to the mentally ill. Up to about 1930 it was generally accepted that their treatment should be institutional, with a maximum of segregation and a minimum of medical care, the medical profession generally not being particularly interested in this type of patient. Since then new methods of treatment (in particular insulin and shock therapy) and new drugs have been introduced, and the emphasis has shifted from custody to cure. The substitution of the new name of mental hospital for the old name of asylum has come to have a real meaning.

Public expression to this new attitude was given in the Report of the Royal Commission on the Law relating to Mental Illness and Mental Deficiency and the subsequent Mental Health Act of 1959, in introducing which the Minister of Health said "one of the main principles we are seeking to pursue is the re-orientation of the mental health services away from institutional care towards care in the community".

This re-orientation was in fact already proceeding with considerable rapidity. During the decade 1950/1960 the number of admissions to mental hospitals nearly doubled (from 56,000 to 95,000). Since there had been little change in the number of mental hospital beds this meant a much greater turnover, and a much larger number of ex-mental patients in the community. Over much the same period the number of new attendances at psychiatric out patient clinics rose from 93,000 to 160,000. With this great increase in the number of mental and ex-mental cases outside the walls of the hospital it became apparent, more particularly from two surveys of the experiences of discharged mental patients, one in Scotland (unpublished) and one in East London,* that the provision for their care in the community was inadequate.

Local authorities already had powers, under Section 28 of the National Health Service Act 1946, and Section 29 of the National Assistance Act, 1948, to provide care and after-care and welfare services for these cases.

Neither Act however, specifically mentioned the mentally sick, the former Act referring to "those suffering from illness or mental deficiency" the latter to those "substantially and permanently handicapped by illness" though Circular 32/51 prescribed those "suffering from neuroses and psychoses and other nervous and mental disorders" as among those to be included in the statutory Register of Handicapped Persons. And in fact very little care or after-care was provided by any local authority for the mentally ill.

The Mental Health Act 1959, was much more specific, Section 6 stating that the purposes for which arrangements are authorised or may be required to be made by a local authority under Section 8 of the 1946 Act for the care or after-care of persons who are or have been suffering from mental disorder shall include the provision of residential accommodation, centres for training and occupation and the appointment of mental welfare officers.

Furthermore, a circular from the Ministry of Health directed all local health authorities to submit their proposals for making arrangements for the prevention of mental disorder and the care and after-care of patients suffering from mental disorders not later than the 1st April, 1960. All authorities duly did, but although schemes for a community mental care and after-care service are thus now on paper, there has as yet been little time to develop them.

The following table gives some indication of the response of the Greater London authorities, though the figures for 1961/62 will doubtless be a good deal more illuminating.

* Enid Mills, Mental Patients and their Relatives in East London, article in "Social Work", April 1961.

TABLE IX.—EXPENDITURE ON MENTAL HEALTH SERVICES PER HEAD OF POPULATION BY AUTHORITIES IN THE GREATER LONDON AREA

	1957–58* s. d.	1960–61* s. d.	% *spent* (1960-61) *on Training centres*	*on other services*
County Boroughs:				
Croydon	1 7	2 1½	50	50
East Ham	2 4	2 4½	52	48
West Ham	2 1	2 6	65	35
Counties:				
London	1 6	2 0½	49	51
Essex	1 5	1 7½	66	34
Hertfordshire	1 1	1 4	65	35
Kent	1 4	1 3	82	18
Middlesex	1 9	1 9	59	41
Surrey	11	1 2	56	44

* 1957-58 figures from Hansard, 4th February, 1959.
1960-61 figures from I.M.T.A. Local Health Services Statistics.

The table shows that all authorities except two had, by the end of the financial year following the passage of the Act, increased their expenditure on their Mental Health Services, some by as much as 30 per cent, and again indicates a higher expenditure by the county boroughs than the counties. It also shows that a high proportion of their expenditure is on the mentally subnormal rather than the mentally sick, since all expenditure on training centres refers to the mentally subnormal, and a substantial, but unascertainable proportion of that on other services. This point is enlarged upon later.

The main requirements of an effective community mental welfare service would appear to be an adequate staff, trained in the new approach to mental illness; a close liaison with the hospital service and general practitioners, and a provision of hostels, training centres and social facilities adequate to the needs of the area.

Of these perhaps the most important is the recruitment and/or training of a new type of mental welfare officer. Before 1958 there were three types of worker in this field—the fully trained psychiatric social worker, who was generally attached to a mental hospital or a child guidance clinic, and rarely to be found in local government service; the duly authorised officer, often a former relieving officer under the Poor Law, whose concern was with the certification of the mental patient, and who had little or nothing to do with after care; and the mental deficiency officer, who *was* concerned with care and after care, but only of the mentally retarded. These three types of workers have now to be integrated to provide a service for all types of mentally afflicted in the community.

The progress in making this sort of provision, and the priorities as viewed by the nine authorities in the area vary widely.

In *Croydon* before 1960 there were five duly authorised officers who were also responsible for visiting mental defectives, one of them being the co-ordinating officer for this purpose. Since then two psychiatric social workers have been appointed and one trainee, with another to be appointed shortly.

Close liaison is maintained with Warlingham Hospital, the mental hospital for the area, and regular group discussion meetings take place between the hospital psychiatric social workers and the mental welfare and health visitor staffs. The scheme submitted includes proposals to provide accommodation in which the hospital and local authority staffs can work together. The hospital operates a day centre in Croydon, and also a weekly afternoon psychiatric club for old people, three evening psychiatric sessions for young patients, and two evening sessions for group psychotherapy.

Arrangements have been made with the Mental Aftercare Association for the reservation of 10 beds in a hostel in Croydon for ex-patients fit to go to work and plans are well advanced for the provision of a 60 bed hostel for the mentally ill no longer in need of hospital care.

The *East Ham* mental health staff consisted before the Act, and still consists of five, all formerly duly authorised officers. Some in-service training has been provided at Runwell, in whose catchment area East Ham is, and at St. Andrews Hospital, Bow. A long term plan for an all purpose rehabilitation centre and for hostels is unlikely to be realised for some years.

There is a therapeutic social club meeting once weekly with an average attendance of about 20. This is run by the patients themselves with the help of the mental welfare officers.

In *West Ham* the recruitment of a team for operating the provisions of the new Act is regarded as a top priority. There is now only one of the old type (ex-duly authorised officer and ex-relieving officer) of mental welfare officer: the other five staff comprise one Psychiatric Social Worker (P.S.W.) ("interchangeable with the two in the child guidance service"—see below) and four trainees, all with social science qualifications. There is a prospect of early recruitment of two more P.S.Ws. and the intention is to work up to a total of not less than 10 and to recruit from men and women having a basic training in social work, with the intention of seconding them to a P.S.W. training on full pay.

Great importance is attached to the preventive side of mental health work, and its treatment in the family setting, the medical officer holding it as established that a high proportion of mental illness has its origin in a maladjusted childhood. The department has a part-time consultant psychiatrist who also acts as psychiatric adviser to the community health service, holding four sessions a week, and a child guidance clinic, with a whole-time secretary, a team of consultant psychiatrists, an educational psychologist and two P.S.Ws.

There is a close liaison with Goodmayes hospital, who have agreed that the aim shall be for the mental welfare officer responsible for admission to follow the case up on discharge.

There was formerly a therapeutic social club in the county borough which has lapsed, and it is the intention to revive it at an early date, when more trained staff is available.

Up to the end of 1960, the mental welfare services in the *County of London* were operated centrally under the direction of a specialist principal medical officer with a staff of one senior P.S.W., four P.S.Ws., 26 duly authorised officers and 37 mental deficiency staff, (a total of 68). They have now been decentralised to the nine county divisions, under the immediate direction of a divisional mental welfare officer responsible to the divisional medical officer, the total staff at the end of 1961 being nine divisional mental welfare officers, nine assistant divisional M.W.Os., 24 senior M.W.Os., 15 M.W.Os., and 27 assistant M.W.Os. Seven of this staff are trained P.S.Ws., and in addition there are another four P.S.W. case workers, increasing the total to 88. A varied and comprehensive programme of in-service training has been undertaken.

One of the first actions of the divisional medical officers after the decentralisation was to establish liaison with the mental hospitals concerned with the divisions.

A new day centre for 40 mentally ill was opened in 1960 in Hackney, and the Council shares in the operation of a new day hospital in Bermondsey established by Cane Hill Hospital by the provision of occupational therapy and some social worker services. The Council also supports the rehabilitation centre in Blackfriars run by the Institute of Social Psychiatry and is undertaking the capital cost of adaptation and equipment of a second centre to be opened by the Institute in Southwark.

Premises have been acquired for adaptation as a hostel for 20 patients discharged from hospital, and a site appropriated for a larger purpose-built hostel.

The mental health services in *Essex* are still operated centrally from Chelmsford—a matter of some dissatisfaction in the Metropolitan Essex Boroughs, to which most of the other personal health services have been delegated. The staff of duly authorised officers (D.A.Os.), also responsible for mental deficiency visiting, comprised 25 at the time of the passing of the Act, mostly ex-relieving officers, under a supervisory D.A.O. with an assistant and these are now designated mental welfare officers, six being located at Romford, three at Hornchurch and five at Walthamstow.

There is an establishment for one P.S.W. but this it has not been possible to fill. There is a scheme for the in-training of assistant mental welfare officers (M.W.Os.), who are subsequently to go on a two year P.S.W. course, but although authority has been obtained for two to go on such courses, none had done so up to the end of 1961. Efforts are also being made to arrange a one year course to existing M.W.Os., (meanwhile they have been attending two-three day courses at local mental hospitals) and to arrange joint use of P.S.Ws., with the mental hospitals.

There are plans for hostels for the mentally ill at Romford and Ilford, but these are still in the early stage.

There is a Goodwill Social Club at Ilford for ex-mental patients to which the Council makes grants, and it also does so in respect of Essex patients received in the East Ham Club, and in the London Centre of the Institute of Social Psychiatry.

Hertfordshire at the time of the passing of the Mental Health Act had a staff of 15 duly authorised officers (who were also welfare officers under the National Assistance Act, and spent only about a quarter of their time on mental health duties) and of eight mental welfare officers, of whom three had diplomas in social science. There was an establishment of eight psychiatric social workers, but none of the posts had been filled, and there was also a part-time consultant psychiatrist.

By the end of 1960 there were 10 Mental Welfare Officers, including two P.S.Ws., one appointed as a senior officer in charge of the work. A part time psychotherapist has also been appointed. In one division in particular certain health visitors have been associated with the development of community health work. In addition the welfare officers in the Welfare Department continued to deal with compulsory removals under the new Act. An establishment of one Mental Welfare Officer to 50,000 population has been approved, which would mean an eventual staff of 16. The Committee has agreed that experienced M.W.Os., holding a social science degree or diploma shall be recommended for secondment on full salary for training as P.S.Ws., and there is a varied programme of in-service training. The M.W.Os. are normally located in the Divisions, of which there are seven. At the end of 1960, 1,071 cases were being visited by M.W.Os., of whom 993 were mentally subnormal, and 78 mentally ill, and it was recognised that the visiting of the mentally ill was very limited and much remained to be done in this field.

One mental welfare officer referred to the increasing number of referrals of mental after care cases, and went on to say "such cases provide very rewarding case-work, but are extremely time-consuming, often needing weekly or fortnightly visits for some time after their discharge from hospital. In practically every case, the patient has remarked how helpful and comforting it is to have someone in whom they can confide their troubles and anxieties, as most of them have felt it impossible to discuss their problems with relatives or friends. Many have said that had they been able to pour out their troubles instead of keeping them to themselves, they feel their disturbances would not have become so acute."*

An early undertaking by the senior worker (who incidentally had previously been senior P.S.W. at a local mental hospital), was the establishment of close contacts with mental hospitals serving the county. A particularly close relationship between the mental hospital and the local authority mental service has been established in one division. No hostels are as yet provided, or in the early planning stage. Two social clubs are run in connection with mental hospitals, one of which, known as a therapeutic social club with a "social therapist" in charge, was started by the Institute of Social Psychiatry in 1949 and taken over by the County Council in 1957.

In *Kent*, before the passing of the Act the care of both the mentally ill and the mentally subnormal was undertaken under a central staff by the 21 district welfare officers with their assistants, save that the care of all female patients and of boys under 16 was normally in the hands of health visitors. This was still the staffing position at the end of 1960, though the County has said it is seeking to appoint two psychiatric social workers. All the staff concerned were given the opportunity of attending a week's course in the mental hospitals serving the area.

There is already one home specially for mildly senile old people, Kent and London being the only authorities in the Greater London area having this particular type of accommodation. Proposals for a second are under consideration. No day centre or club is provided by the county, but financial assistance is given to voluntary organisations. There is one notable example of a therapeutic social club the "Stepping Stones Club" in Bromley, but this is a voluntary enterprise, financially supported by the King Edward's Hospital Fund and the local Hospital Management Committee.

Before the passing of the Mental Health Act, the mental health services in *Middlesex* were in charge of a principal

* Report of County Medical Officer of Health, 1960, p. 64.

medical officer asssisted by a senior medical officer, and were organised in five divisions to correspond with the catchment area of the mental hospitals. There were 26 duly authorised officers, five mental health social workers and eight psychiatric social workers.

The organisation is now still the same, but an additional senior medical officer has now been appointed, Mental Welfare Officers (formerly D.A.Os.,) now number 30, and a post of County Psychiatric Social Work Organiser has been created and filled.

In his first report, the County Psychiatric Social Work Organiser says "As no effective case work can be done in the community without the experience and guidance of the hospital staff, I tried to build bridges where there were none, and build stronger ones where there were already some. This co-operation and co-ordination with the hospital I consider one of my most important tasks . . . The social workers in hospitals and I had useful discussions as to the standards of social work in the community and we tried to minimise the duplication of work."

He also recorded having arranged with the Extra-mural Department of London University a course of 30 sessions in human relations for staff (15 on social work theory and 15 seminars) spread over an academic year—"for some, to bring them up-to-date, for others, particularly the new entrants to the field, to lay down the formulation of professional social work practice."

Plans for the provision of two hostels for the mentally ill, one purpose built and one an adaptation, are in active preparation but they are unlikely to be open until 1963. Meanwhile an experimental unit has been brought into use for six selected women ex-patients where they will care for themselves without residental staff. Three social therapeutic clubs are run by the county, the total attendance in 1960 being 1,763. There were also over 4,000 attendances at the day clubs and day rehabilitation centres run by the Institute of Social Psychiatry in respect of which the Council make payment.

Before 1960, *Surrey* employed on their mental health services, under an administrative officer, 15 mental welfare officers plus two specialist health visitors to the mentally deficient, while the routine visiting of women and girls and boys under 16 was carried out by health visitors.

There are now 27 mental welfare officers and six specially trained health visitors, and a specialist medical officer and a Psychiatric Social Worker have joined the headquarters administrative team. These two have especially concentrated on liaison with the mental hospitals, of which there are six in the county and the six specially trained health visitors have been seconded to these hospitals to undertake community care work from them.

In-service training is being given, and it is planned to recruit a further four social work trained officers, with a view to their qualifying as P.S.Ws., later.

An extensive provision of hostel accommodation is planned, but none is yet open, though it is hoped that one for 18 residents in the grounds of Netherne Hospital will be available by the end of 1962. All hostels, whether for the mentally ill or for the subnormal, will be run by the Health and not the Welfare Department.

An evening social club is staffed by Surrey in connection with Epsom Hospital, and it is hoped to develop this into a day centre later.

It will be seen from the foregoing pages that the sense of urgency with which the questions of recruitment and training for the community health services has been tackled varies widely among the authorities in the Greater London area.

Appreciation by local authorities of the need for co-operation with the mental hospitals appears pretty general, but it has been more successfully achieved by some than by others. Liaison appears to be particularly effective in Middlesex, Croydon and West Ham, and it is noteworthy that here the local authority areas correspond with the hospital catchment areas. It is much to be hoped that if the Government's proposals for reorganisation are implemented some consideration will be given to the relationship of these catchment areas to the new London Borough boundaries, especially in face of the view recently expressed in a *Times* special article (15th August, 1962) of the need for "the closest co-operation between the hospital authorities and the mental health department of the local authority, if not their virtual integration." This co-operation is particularly difficult to effect in the central London area, owing to the general location of the mental hospitals on the periphery outside the present county boundary.

The prospect of securing any coincidence between the catchment areas of *general* hospitals and local authority boundaries seems hopeless on account of the principle of freedom of choice of hospital. But mental hospital catchment areas are already based on borough boundaries, and in the London area there are only three boroughs (Southwark, Lambeth and Lewisham), divided between two regional hospital boards, and only three more (Kensington, Westminster and Hammersmith) in the catchment area of more than one mental hospital. But unless some action is taken, many of the new boroughs will be in more than one area (the union of Acton, Southall and Ealing for example will be in three) so greatly complicating the problem of co-operation.

The *Times* article quoted above goes on to say "the overcrowding . . . typical of many mental hospitals could be solved overnight if there was sufficient geriatric and hostel accommodation."

Only one or two hostels have yet been provided for discharged mental patients, and although some authorities plan them, others do not. Only two residential homes for mildly senile old people are provided, by Kent (who propose to provide a second) and by London, both under the National Assistance Act, though there are numbers of senile in Part III accommodation, especially in the former institutions.

The situation here gives an indication of how closely interdependent the welfare services are. The mental hospitals are less efficient than they could be because of overcrowding. Many of their patients could be suitably accommodated in hostels or Part III accommodation. But hostels are practically non-existent, and there are waiting lists for all local authority old people's homes. Some of those in these homes however, might well be in homes of their own if they were available but they are not. Hostels and homes can be built under the Mental Health Act, the National Assistance Act and the Housing Acts, housing for the old under the Housing Acts. But as yet a solution of this accommodation problem has nowhere been comprehensively and co-operatively planned by the departments and authorities involved. There is however, some hope that it now will be as a result of Ministry of Health Circular 2/62 requiring county and county borough councils to submit 10 year plans for their Health and Welfare Services by Oct. 31st, 1962.

II. THE MENTALLY SUBNORMAL

The change of attitude towards the mentally subnormal, and the advances of medical science in their treatment have been less dramatic than in the case of the mentally ill. This is hardly surprising since mental deficiency is in the main a static, and incurable condition, and whereas annual discharges from mental hospitals number about 60 per cent of their population, those from mental deficiency hospitals number only about 6 per cent.* This is not to say no progress has been made. There is now general recognition that adverse environment may have a marked effect in preventing the full development of limited mental powers, and that the mentally subnormal (those with I.Q's from 50-70) and even some of the severely subnormal (I.Q's below 50) have greater capacities and respond more readily to psychological and material incentives than was at one time thought. There is moreover less implicit reliance on the validity of intelligence tests. All this implies that more of the mentally subnormal are capable, with assistance, of remaining in the community and even of making a contribution to it, than was once considered possible.

At the same time, while the number of discharges from mental deficiency hospitals has been increasing, the demand for beds is increasing too, and there are now waiting lists all over the country. The steepest increase is among the middle aged and elderly. More now survive to a later age, and are

* PEP, Mental Deficiency and Social Care, quoting from Registrar General's Statistical Review of England and Wales, 1957-58.

admitted as a result of the death or infirmity of those caring for them. This again emphasises the need for a greater provision of community care where this is practicable.

Local authorities have had a responsibility for the mentally subnormal ever since the Mental Deficiency Act of 1913 (now repealed) which gave them the duty to ascertain defectives (classified as idiots, imbeciles, feeble-minded and moral imbeciles) who were "subject to be dealt with", to place them under supervision and, if considered necessary, to secure their admission to institutions under a prescribed procedure. The emphasis of the Act, in accordance with the sentiment of the time, was on the protection of the community against the defective rather than the welfare of the latter, and the supervision was evidently meant to be directed rather to seeing whether he was safe to remain at large than securing his well-being. It was not until 1927 that an amendment of the 1913 Act required local authorities to provide "suitable training or occupation" for those under supervision, while the National Health Service Act 1946, first introduced the conception of "care and aftercare". The Mental Deficiency Service like the Blind Welfare Service, thus has a fairly long history and in consequence is also more uniform in coverage by the various local authorities than some of the more recent services, as is shown by the following figures:—

TABLE X.—PROVISION FOR THE MENTALLY RETARDED BY AUTHORITIES IN THE GREATER LONDON AREA*

Authority	*No. of Training Centres*	*No. of places*	*No. under Supervision*	*No. of Home Teachers*	*Waiting list for hospital accommodation*
County Boroughs					
Croydon	1	60	482	None	10
East Ham	1	50	277	"	10
West Ham	2	120	182	See Text	6
Counties					
London	21	1,335	4,389	3	230
Essex	12	794	3,587	None	356
Herts.	6	250	993	1†	43
Kent	11	560	1,627	8	192
Middlesex	12	873	3,042	1†	156
Surrey	8	425	1,458	None	126

* Figures extracted from the latest available Medical Officers of Health Reports.
† Recently provided on a limited and experimental basis.

has published figures showing the proportionate training centre provision by local authorities in England and Wales. The following relate to the London area.

Authority	*No. of children and adults receiving training per* 100,000 *population—31st December,* 1959
East Ham West Ham	50–59
Essex Middlesex	40–49
Croydon Kent London	30–39
Herts. Surrey	20–29

A recent estimate of the number of severely subnormal per 100,000 population in England and Wales is 187. At the same date, these authorities' estimates of the percentage provision made to that required, based on their own assessments of suitability, for training was:—

90–99%	Kent, London, Surrey
80–89%	East Ham, Middlesex
70–79%	Essex
60–69%	Croydon, West Ham, Herts

Among other interpretations which might be given to this table on comparing it with the previous one is that it represents an index of complacency. Perhaps because it is a well established service, proposals by the local authorities for developments for the mentally sub-normal are generally more specific than for the mentally ill. Staffing and training proposals, discussed above, cover both classes, since this is to be a unified service. Otherwise the main provision contemplated is more and better training centres for which there is plenty of scope. Purpose built centres are to replace those in church halls, etc. Separate and appropriate provision is to be made for adults and juniors where this is now shared. Most authorities have already made a start with this new type of provision. In West Ham, for example, where there was formerly only one centre there are now two, one for adults and one for juniors, and plans for a purpose built junior centre are well advanced. There is an increasing trend, more marked at present in London and Middlesex than elsewhere, to give the work in the adult centres an industrial rather than a purely occupational slant. In Middlesex a particularly interesting experiment has been successfully developed by which a number of subnormal youths are actually employed in a factory, their wages at the negotiated piece work rates being paid over to the Council. Three of the youths have now been given permanent employment in the factory. In London there are two specifically industrial training units and industrial out-work is undertaken at all the adult centres, an industrial manager having been appointed. Surrey's one adult centre is an industrial centre, and it is planned eventually to have as many adult as junior centres, all of them industrial.

Croydon's plans for a joint workshop, with an industrial slant, for both the mentally and physically handicapped, is of particular interest in view of the views on adult training centres expressed by Dr. Chalmers Keddie.*

"Should the centre be exclusively used by the mentally handicapped? In my view no. It may be all right for a large authority to build a centre just for the mentally handicapped, but I feel there is tremendous scope here, especially for the smaller authority, to have a centre which will take all their handicapped. Not only the mentally handicapped, not only psychiatric cases, but also physically handicapped. Each individual would have some special skill that could be developed and the various personalities would influence each other and the centre would benefit."

London and Surrey are the only authorities which already have hostels for the subnormal. London has one for girls discharged from E.S.N. schools and one for youths, and five more are planned to be opened by 1965. Surrey has one for acutely subnormal children. Middlesex has no hostel of its own, though it is hoped to have one by 1963 and more eventually, but it

* "Community Care of the Handicapped". National Society for Mentally Handicapped Children, 1960.

makes use of one for each sex provided in the county by the Mental Aftercare Association. The other authorities have none; Essex plans three by 1963, Hertfordshire two and Croydon one. Kent, East and West Ham have no specific plans as yet. On the voluntary side, apart from the Mental Aftercare Association homes, the National Association for Mental Health with the aid of a grant of £50,000 from the City Parochial Foundation, has started two hostels, one for boys in 1958 and one for girls in 1960. These hostels were intended to be experimental and one reason why local authorities are slow to build is that there is not yet sufficient experience of the amount and type of accommodation required. A feeling is growing that this should be governed by the social needs of the resident rather then their medical categories.

"Special care units" are very few throughout the country, and Middlesex and London are the only authorities in the London area so far providing them. To quote the report of the Medical Officer of Health of Middlesex for 1960: "These units cater mainly for low grade children (many of whom are also physically handicapped) who are capable of receiving only care and basic habit training, and those children of higher intelligence whose overall performance and behaviour is of very low level because of maladjustment. The removal of these children from the family home during the day greatly eases the emotional tension within the family circle and in such cases special guidance and help has been given to the families with remarkable improvement in the behaviour and performance of the children in the special care units within a comparatively short time. . . .With the continued shortage of beds in psychiatric hospitals it can be said that special care units are successfully filling a need that cannot otherwise be provided, especially as the children concerned are likely to suffer progressive deterioration in their condition, unless some provision is made for their care and training." Because of their success, Middlesex proposes to provide these units in all eight of its junior training centres (three are operating at present) and to increase the number of places in them from 12 to 24. Surrey has none at present but plans in due course to have one in every junior centre. London has one centre now functioning in a new purpose built junior centre, but plans to provide one in each such centre in future. West Ham is incorporating a special care unit in a new purpose built junior training centre, which it is hoped may be completed by 1964. London has also experimented in the provision of play centres for severely subnormal children during the holiday season.

There are also nine "community centres" functioning in Middlesex where parents and others can obtain advice and medical and psychiatric assistance, and West Ham has a similar type of provision in its "development clinic" where parents whose children through mental or physical defect are abnormally slow to development are advised on their upbringing.

Kent is the only authority making provision for home teaching of the mentally subnormal on a substantial scale. West Ham has an occupational therapist who does a certain amount of home instruction, in addition to running a rehabilitation centre attended by the mentally affected as well as the elderly recovering from strokes, and some physically handicapped. This may eventually be absorbed into the adult training centre, and an attempt is being made to get industrial contract work in connection with it.

In view of the fact that it is estimated that more than half of the severely subnormal are retained by their parents at home, and further, that because of the pressure on the hospitals even urgent cases, whose relatives would wish to see them in institutional care, may have to remain at home for long periods, it would seem that one of the most valuable services the local authority can offer is to give support to these families. The burden of anxiety and the dislocation of the normal pattern of family life which such cases can impose, and the tensions which may be set up in a family as a result of their presence, can be tragic in their effects. Special care units, of which so few exist or are planned at present, and consultative clinics can do much to relieve the burden. So also can the provision of holidays. The National Association for Mental Health provides some holiday homes, but it appears that as yet only East Ham and

Middlesex of the local authorities are themselves organising holiday parties on a substantial scale.

There also seems considerable scope for the extension of industrial training and placement schemes for the adult subnormal. There is a good deal of evidence that once they are trained in the sort of repetitive process characteristic of much of modern light industry, their output can equal that of the normal worker, the result being that they can become self-supporting, as well as having the satisfaction of becoming occupied and useful.

With regard to the mentally affected generally, the following extract from the article by Miss Enid Mills, mentioned on p. 59, although it refers specifically to the mentally ill, gives some indication of the scope for reorganisation which is offered to the new London Boroughs in respect of this service, and indeed of the welfare services generally:

"Fragmentation in the services offered to these people has arisen mainly because of specialist interests in the same areas. London presents an immense problem to anyone wishing to rationalise the services available to any group. To do this for the mentally ill it would be necessary not only to integrate the services specifically set up to deal with mental illness, but also to persuade other voluntary and statutory bodies dealing with various aspects of the patients' lives to take an overall view of the needs of the mentally ill."

CHAPTER FIVE

Services for the Socially Handicapped

I. DEPRIVED CHILDREN

LIKE those for the blind and the mentally retarded, the services provided by local authorities for deprived children have a pretty definite scope, and their administration in the Greater London area is comparatively uniform. While the uniformity of the first two is partly due to the fact that they are long established services, the service for deprived children is comparatively new, but it has taken shape on the lines laid down in the Curtis Report,* and the Home Office, since the local authorities' Children Departments were established under the Children's Act of 1948, has extended closer control over them than has the Ministry of Health over Health and Welfare Departments. In particular strong emphasis has, from the Curtis Report onwards, been laid on the importance of providing for deprived children a life approximating as closely as possible to that which they would lead in a normal home. It is generally agreed that this can best be done either by boarding out in foster homes, or by accommodating the children in very small homes with a housemother or houseparents so that a family atmosphere can be established. In particular, stress is now laid on the importance of individual care for the very young child—i.e. of nursery age.

The success of the authorities in the Greater London area in realising these methods of disposal are shown in the two tables on p. 77.

The most conspicuous feature of these tables is the success of the county boroughs both in keeping their homes small and in boarding children out. No nursery accommodates more than

* Report of the Care of Children Committee 1946, Cmd. 6922.

TABLE XI.—AUTHORITIES IN THE GREATER LONDON AREA: NO. OF CHILDREN IN CARE MARCH 1961, WITH PERCENTAGE BOARDED OUT

Authority	*Total number in care*	*No. in care per 1,000 of estimated population under 18*	*Percentage boarded out*
County Boroughs:			
Croydon	294	4·5	56
East Ham	177	7·1	67
West Ham	208	4·8	40
Counties:			
London	8,884	11·6	23
Essex	1,582	3·1	60
Hertfordshire	950	3·9	63
Kent	2,361	5·2	41
Middlesex	2,701	5·0	35
Surrey	1,383	3·7	44
England & Wales	62,199	5·0	48
County Councils (England & Wales)	41,713	4·8	46
County Boroughs (England & Wales)	20,486	5·5	52

TABLE XII.—AUTHORITIES IN THE GREATER LONDON AREA: SIZE OF ACCOMMODATION IN CHILDREN'S HOMES AND NURSERIES

Nurseries

Authority	25 *beds and under*	26–50	*over* 50	*Total*
County Boroughs:				
Croydon	1	—	—	1
East Ham	1	—	—	1
West Ham	1	—	—	1
Counties:				
London	11	13	5	29
Essex	3	—	—	3
Hertfordshire	1	1	—	2
Kent	4	1	—	5
Middlesex	7	4	—	11
Surrey	11	—	—	11

Homes: including reception homes: excluding remand homes and hostels

	15 *and under*	16–25	26–50	51–100	*over* 100	*Total*
County Boroughs:						
Croydon	7	—	—	—	—	7
East Ham	6	—	—	—	—	6
West Ham	9	—	—	—	—	9
Counties:						
London	98	5	2	1	6	112
Essex	15	10	—	—	—	25
Hertfordshire	1	3	3	—	—	7
Kent	10	4	2	3	—	19
Middlesex	60	3	3	—	—	66
Surrey	28		—	—	—	28

25 children, no home more than 15. Two of the three, as against two of the six counties, have a percentage boarded out well

above the national average. The third, West Ham, while below it, has a particular problem with large numbers of Irish in the dock areas, and the difficulty of boarding out Roman Catholic children as compared with finding them vacancies in homes.

Moreover, the attached figures show that over the past few years the county boroughs have achieved a greater increase in the percentage boarded out than the counties.

TABLE XIII.—AUTHORITIES IN THE GREATER LONDON AREA: PERCENTAGE OF CHILDREN BOARDED OUT, MARCH 1956–61

	1961	1960	1959	1958	1957	1956
County Boroughs:						
Croydon	56	58	53	52	52	55
East Ham	67	68	53	49	42	41
West Ham	40	28	33	33	30	32
Counties:						
London	23	23	21	22	22	22
Essex	60	60	60	59	60	60
Hertfordshire	63	66	66	68	68	67
Kent	41	43	43	41	44	44
Middlesex	35	36	38	35	37	37
Surrey	44	46	45	43	44	44

The East Ham figures are particularly notable and show that a concentrated effort can steadily increase the numbers boarded out even in a mainly one class built-up borough. And at the present time 35 per cent of those boarded out in East Ham are inside the borough boundaries, as against 65 per cent outside.

London, on the other hand, has the lowest proportion boarded out in the country, and there are only five other authorities with a lower proportion than Middlesex. London again is the only authority in the area with children's homes accommodating more than 100, and it has six of these, four with more than 350. Whatever the efforts made to break up these large homes into smaller units on the cottage home principle, they must still remain artificial children's colonies.

London, with 8,800 children in care in 1961 and more than 9,000 today has, of course, a huge problem. Middlesex came next with 2,700. Only two other authorities (Kent and West Riding) had more than 2,000 in care, and only eight others, (including Surrey and Essex) more than 1,000. Strenuous efforts have been made to cope with these large numbers in

London and Middlesex, and London's provision of over 90 family sized homes since 1948, and Middlesex's of 60 are notable achievements. But the figures indicate that it is easier for a small authority to produce an approximation to normal home life for its children in care than a larger one.

Against this disadvantage it is claimed that the large autho- ity, with its larger catchment area, can make provision for small groups with special needs in a way impossible for a smaller authority, where the very limited number in any particular category would make special provision totally ι neconomic.

There are two answers to this contention: first, the case for the provision of special residential accommodation for the abnormal child (including the maladjusted, the educationally subnormal and the physically handicapped) can be over stated. The Children's Officer for Kent,* in commenting on the dangers of impersonality in such establishments has said: "There is no doubt that the ideal provision for a school-age handicapped child in care would be for him to live in a foster home or small Children's Home or attend a day Special School." This approach, regarding the handicapped child as a child, needing warm human contacts, may normally be expected to have happier results than regarding him as a specimen to be segregated with other specimens to facilitate a clinical approach. The same sort of criticism can be levelled at the large reception home, where the child, having just been uprooted from familiar surroundings may undergo a series of tests and examinations before being allocated to what is regarded as the most suitable type of provision, where he then has to make a fresh readjustment. There may be something to be said for the practice of the smaller authority of an experienced child worker making a spot assessment and placing the child in its permanent home, if it stays, with a subsequent change of disposal in the case where the assessment proves wrong.

Secondly, where specialised establishments are found to be essential, they can be and are provided by one authority

* Report of the Work of the Children's Committee, 1953-58, Kent County Council.

accepting cases from others, and by voluntary organisations supported by per capita payments or subsidies from local authorities. With all its resources, the London County Council, for example, has no approved school for girls, and although it has its own remand home for girls, it also makes use of that maintained by Croydon.

While the main responsibility of Children's Departments is the provision of care for children who are deprived of a normal home life, they also have a responsibility for those placed under their care by the courts. Of 45,203 children received into care in England and Wales in the year ending 31st March, 1961, 4,112 were committed into care under "fit person" orders made by the Juvenile Courts. The proportion is higher in London, and this type of case normally remains longer in care, so that of 9,198 children in care in London in August 1961, 2,642 were from the Juvenile Courts. This has meant that a close relationship has been built up between the courts and officers of the Children's Department of the London County Council, and considerable consternation has been expressed at the prospect of the disappearance of the Council. In particular, it has been urged that a child may commit an offence in one London borough, live in another, and appear in a court in a third, and that without one central authority the problem of representation at the courts would be hopelessly complicated. Here the basic difficulty is in the fact that the areas of jurisdiction of the Metropolitan Courts bear no relation to borough boundaries. The Interdepartmental Committee on Magistrates' Courts in London has now made its recommendations (January 1962), and the Government has expressed its intention of acting on them in the next session. These recommendations advocate that the areas covered by the courts, including Juvenile Courts, be redrawn on the lines of the borough boundaries, with due regard to the Government proposals for the new London Boroughs, and also that Metropolitan Magistrates should have jurisdiction over offences arising anywhere in the area covered by the integrated courts and should be able to exercise it in any court in that area.

This means that there should be a much closer relationship

between the Children's Departments of the new London Boroughs and their respective Juvenile Courts than has ever been possible under present arrangements, where the nine divisions of the L.C.C. Children's Department have no correspondence with the areas of the eight Juvenile Courts.

London also has close relations with the courts by reason of its three approved schools for boys and two remand homes (all the other counties except Kent also have approved schools, but none of the county boroughs, and all the counties (except Hertfordshire) and one county borough have remand homes). In particular at its boys' remand home London runs a classification centre for approved schools, an undertaking which it is justly claimed could only be assumed by a large authority. Who then would run it if the London County Council disappears? Here it may be asked whether responsibilities in connection with the administration of justice are properly the function of local government. There are at present 117 approved schools in England and Wales, only 26 of them run by local authorities, the rest by voluntary managers. All other penal and reformatory establishments are under central government, as also is the probation service. There is notorious difficulty at the present time in obtaining vacancies both in approved schools and remand homes, and there seems a strong case for placing these also under direct Home Office control.

TABLE XIV.—Illegitimate Births in the Areas of Authorities in Greater London

Authority	*Annual Total of illeg. live births*	*% of illeg. births to total live births*
County Boroughs:		
Croydon	288	7·2
East Ham	75	4·9
West Ham	144	5·3
Counties:		
London	6,530	11·4
Essex	1,056	3·6
Hertfordshire	543	3·7
Kent	1,224	4·6
Middlesex	2,181	6·2
Surrey	949	4·4

II. THE UNMARRIED MOTHER

The size of the problem with which any authority is faced in dealing with the unmarried mother naturally varies with the number and proportion of illegitimate births in the area. These vary considerably as is shown by the figures (Table XIV) extracted from the latest available reports of the respective Medical Officers of Health.

The figure for London is outstanding, and is commented on as follows by the Medical Officer of Health in his report for 1960:

"The percentage of illegitimate births in London has shown a marked upward movement over the last five years whilst nationally there has been a relatively slight increase. Boroughs with particularly high percentages were Paddington (21·9), Kensington (16·8), Lambeth (15·1), Chelsea (14·1) and St. Pancras (14·0). The only other large town in England and Wales with an illegitimacy rate approaching that of London is Manchester, where there has been a rise from 6·6 in 1951 to 9·3 per cent in 1959.

A complex of factors probably accounts for the phenomenal rise in London—proportionately more single women, a continuous influx of unmarried women, many of whom are already pregnant, and the facilities which London can offer

UNMARRIED MOTHERS SEEN BY MORAL WELFARE ASSOCIATIONS IN LONDON, 1ST OCTOBER, 1959—30TH SEPTEMBER, 1960

(Figures in brackets are for 1959)

	British (U.K.)	*Eire*	*European*	*West Indian*	*Other*	*Total*
Non-Londoners pregnant on arrival in London	424 (383)	272 (183)	50 (39)	85 (64)	33 (19)	864 (688)
*Non-Londoners not pregnant on arrival in London	71 (93)	87 (96)	8 (17)	33 (20)	3 (6)	202 (232)
Resident in London one year or more	1,371 (1,270)	454 (455)	76 (85)	267 (236)	58 (39)	2,226 (2,085)
	1,866 (1,746)	813 (734)	134 (141)	385 (320)	94 (64)	3,292 (3,005)

* Had lived in London less than 12 months before making contact with moral welfare association.

to an unmarried mother in the way of anonymity, ante-natal care and support from moral welfare organisations. Information as to the extent of some of these factors will only become reliably known when the results of the 1961 census are available.

The table above gives details of women seen by the moral welfare organisations in the twelve months ended September, 1960, from which it will be seen that 864 (26·2 per cent) were pregnant on arrival in London and that in all 1,426 (43·3 per cent) were not British. It should be remembered that these components of the illegitimate births are minimal—the moral welfare organisations do not deal with all unmarried mothers, though doubtless they will tend to deal with proportionately more of the non-Londoners."

As is hinted by the reference to moral welfare organisations above, moral welfare work is the one remaining welfare service where the bulk of the work is still undertaken by voluntary bodies although, following Ministry of Health Circular 2866 of 1943, local authorities now play a larger part than they did before the war. These are largely run on a religious basis through the Diocesan Councils for Moral Welfare and the Local Moral Welfare Associations of the Church of England, but there are similar Roman Catholic, Free Church and Jewish organisations. Policy is influenced by the National Council for the Unmarried Mother and her Child, and nearly all moral welfare workers who staff the local associations are trained at the Josephine Butler Memorial House.

Initially, practically all prospective unmarried mothers requiring help are referred to appropriate moral welfare workers by maternity clinics, general practitioners, social workers, clergy and others, for advice and any necessary action. Such action frequently includes arranging admission to a mother and baby home, and the great majority of these homes are again maintained by voluntary organisations, the figures in 1956 being 27 local authority as against 107 voluntary homes.*

* Virginia Wimperis, "The Unmarried Mother and Her Child", George Allen & Unwin, 1960.

In London for example, of the 3-4,000 unmarried mothers seen by the moral welfare workers annually, about 800 are received into voluntary mother and baby homes where they normally stay six weeks before and six weeks after confinement, the Council paying an agreed maintenance rate. Rather fewer are received without payment from the Council and another 300-400 are received into the Council's own establishments, but some of these are not seen by the moral welfare workers. The moral welfare workers normally help those who do not enter homes with their problems, and co-operate with the home staff in the after care (including assistance with adoption) of those that do, and the moral welfare associations, as well as the voluntary homes, are financially supported by the local authorities.

Croydon, East Ham, West Ham, and Essex all undertake the care of the unmarried mothers almost entirely through the local moral welfare associations, liaison being effected through the health visiting service. Kent, Hertfordshire and Surrey do so in the main but Kent and Surrey each has one mother and baby home of its own (Kent with 22 beds, Surrey, 12), and Hertfordshire has a special almoner for dealing with these cases.

Middlesex, while also making use of voluntary homes, has four homes of its own, with 79 beds, and the care and aftercare of its cases are in the hands of a special service almoner with two whole time assistant almoners.

In all these authorities, work in connection with unmarried mothers is undertaken in the Public Health Department, generally under the care of the superintendent health visitor. In London however, it is mainly the responsibility of the Welfare Department. A social worker with an assistant arranges with the moral welfare workers for the reception of cases into the voluntary mother baby homes as the responsibility of the Council, and the Department also has three establishments of its own—Carisbrooke house, with beds for 24 women, and two units in large homes, one for 18, the other for 12, temporarily 25. The social work on these cases is undertaken by social workers in the department in co-operation with the matrons and the moral welfare workers concerned. On 31st March, 1962, there were 177 unmarried or expectant mothers and 94 babies in voluntary homes as the responsibility of the Council, and 64 mothers and 53 babies in the Council's own establishments. These are those presenting the more intractable problems, the voluntary homes generally limiting themselves to mothers of good character with their first baby, the few taking second baby cases being very selective.

In 1944 the Welfare Department was instrumental in establishing a Joint Advisory Body of official and voluntary bodies dealing with the unmarried mother in the county, but this body is now administered by the Public Health Department. Early in its life it recommended the establishment of a hostel for mothers who had no home to go to but wished to retain their babies and go out to work. As no voluntary organisation felt able to undertake this responsibility, the Welfare Department did so, and it still maintains this hostel, which accommodates 25 mothers with their babies. A similar hostel subsequently opened by a voluntary body closed for lack of funds, (this type of case requires considerable subvention) and was subsequently taken over by the Middlesex County Council as an ordinary mother and baby home. Wadham Gardens, the L.C.C. Hostel, is thus the only one of its type in the London area.

This is one illustration of the fact that, while the provision made for the unmarried mother and her child during the antenatal, and immediate post-natal period is fairly complete, the provision for their welfare after that time leaves much to be desired, more especially in the cases where the mother keeps her child and is sometimes faced with almost insuperable problems. To quote Virginia Wimperis* "It is a sad feature of the case work among unmarried mothers at present that . . . in the vast majority of cases moral welfare workers lose touch soon after the mother leaves a Home to take her first job, or when an affiliation or an agreement with the child's father is safely negotiated . . . They lose touch chiefly because they are overpressed . . . This question of long-term after care indeed needs consideration but since any extension of case-work would

* Op. cit. pp. 208-9.

require further funds it again raises the question of why, in a field so responsive to the personal touch, voluntary societies that are so experienced and so well equipped to give help are so disastrously short of funds."

III. PROBLEM FAMILIES

Problem families, though the term has its critics, are a pretty generally recognised social phenomenon. Various definitions have been offered, among them "families who, because of personal inadequacy and adverse circumstance, have lost the will power, and perhaps even the wish, to maintain normal standards of home and child-care"—"families who manage their household affairs in so unsatisfactory a manner as to present a problem, or a series of problems, for various public services"—"households where dirt, disintegration and disorder are present to such an extent that the social habits of the family stand out in contrast in the community in which they live."

Practically all the authorities in the London area have made some sort of special provision for this type of family, but the scope and nature of the provision, and the department responsible for making it, show considerable variation. The families may become apparent to the Public Health Department, through the health visitors or public health inspectors; to the Children's Department, through applications for care or complaints of neglect; to the Welfare Department, because of homelessness since failure to pay rent, a frequent symptom in these cases, leads to eviction. Once in temporary accommodation, these families are exceedingly difficult to rehouse, since landlords municipal or otherwise, are unwilling to accept them. It is this factor in particular which has led some welfare departments to attempt to better these families rather than have them on their hands indefinitely (see also Chapter Six).

As stated in that chapter, *Croydon* initiated a special welfare service for homeless families following the imposition of more stringent conditions for admission. The health visitor seconded to the Welfare Department to assist the homeless families to reintegrate themselves into the community also has referred to her by other departments of the council and by outside bodies any family in danger of becoming homeless, including problem families. Thus in 1960 she had referred to her 46 families admitted to temporary accommodation (of whom only four were without permanent accommodation at the end of the year), 43 potential homeless families and 53 problem families.

In *East Ham* there are no special arrangements for dealing with problem families, this being regarded as one of the functions of the health visitors, in respect of which the Medical Officer of Health comments rather illuminatingly as follows: "It is the problem family which makes the greatest demands on the time and resources of the health visitor. Access to such homes is notoriously difficult, accurate information as to income budgeting and domestic circumstance only elicited with almost surgical dexterity and time seems to have little importance here . . . Indeed it can rightly be said that the 'problem family' is the most exhausting and least rewarding of all the tasks the health visitor is asked to deal with". This seems to imply that qualities are called for which can hardly be expected in every health visitor.

Cases in the county borough where the homeless cannot be rehoused are rare, but when this occurs the children are taken into care. In this connection it must be remembered that East Ham is the smallest of the county boroughs in Greater London and that admissions to temporary accommodation have averaged fewer than 50 a year over the past few years.

West Ham has since 1958 had a problem family case committee, consisting of the Senior Medical Officer, superintendent health visitor, three other health visitors, and the superintendent nursing officer as permanent members, with others, such as the home help organiser, psychiatric social workers, officers of the children and housing department, and the N.S.P.C.C. inspector and social worker, attending as desirable. The object of the committee is to give support to the officer chiefly concerned with any problem family, and it is regarded as complementary to the "well established Officers' co-ordinating Committee". These co-ordinating committees function in all authorities in the Greater London area, having been established following the joint circular of the Home Office and Ministries of Health and Education in 1950 on the neglect and ill-treatment of children in their own homes and that of the Ministry of Health in 1954 on avoiding the break-up of families.

In 1961 a joint committee consisting of the Town Clerk, Borough Treasurer, Minister of Housing, Housing Manager, Chief Welfare Officer and Children's Officer on "Unsatisfactory Tenants and Problem Families" made the following recommendations:

A. That the Council accept as a matter of policy that families which are unable to manage their household affairs to such an extent as to require assistance from Corporation Departments (in this report called "Problem Families") should be identified at the earliest possible moment, and that concentrated efforts should be made to enable them to become fully responsible for their own affairs.

B. That the functions of the Officers' Co-ordinating Committee for Neglected Children be extended to enable it to deal with all appropriate problem families.

C. That the Housing Committee be asked to accept the responsibility of providing suitable accommodation for families in need of preventive supportive or rehabilitatory work which are either evicted or homeless or whose living conditions jeopardise their successful progress towards full responsibility.

It will be noted that these recommendations stress the importance of housing in relation to problem families; and the last recommendation is more practicable for an authority responsible for housing as well as the other welfare services.

In *London* the Welfare Department, the Public Health Department and the Children's Department all have arrangements for dealing with problem families.

With the very large numbers of the homeless coming into temporary accommodation in the early 'fifties, it soon became apparent in the Welfare Department that among them were families with no hope of acceptance by a housing authority on account of their past records, and who therefore might be expected to remain indefinitely unless some effort were made to rehabilitate them. Accordingly two so-called "special units" were set up in 1953 for this purpose. The particular feature of these units was they accommodated fathers as well as mothers and children (some voluntary homes of this type do not); each family had its own rooms, and its own cooker and sink, and the warden and assistant warden took alternate 24-hour shifts, so that there was always responsible supervision at critical times of the day—in the early morning, getting husband off to work and children to school, in the evening, getting the husband his meals and the children to bed.

The scheme met with a large measure of success, the great majority of those received being eventually successfully re-housed, and two further units were opened by 1957 making a provision for a total of some 25 families. Recently less use has been made of them—the number of families in March 1962 was down to 15, and one of the units was closed to problem families during the year and used for the ordinary homeless. This suggests that they may now be regarded as less successful than initially, since the experience of the Public Health Department (see below) does not indicate that the number of problem families has decreased. Certainly recruitment of wardens is not easy, both because of the personal qualities required and the stamina demanded.

It should be borne in mind that these special units cannot take in families with any sort of home of their own, however deplorable, and however effective such a change of environment might be thought to be. They have however, received and so reunited a limited number of families where the children were in care following a break-up, generally due to homelessness.

In the Public Health Department there are now 13 full time social case workers in the nine divisions undertaking the support and intensive care of problem families, assisted part-time by another 15 workers who are also employed as treatment organisers, the case workers undertaking 9-10 cases at a time the treatment organisers 3-4. Special assistance is also given to problem families by selected home helps.

The Children's Department too has appointed caseworkers to deal with problem families, but the policy now is rather for each child care officer to deal with her own cases. Close liaison is maintained with the Public Health Department through the Co-ordinating Committees in the Divisions, where there are periodic policy meetings and fortnightly case conferences. These are also attended by representatives of the Welfare, Housing and Education Departments, with representatives of other bodies co-opted as necessary.

The London County Council also grant-aids Family Service Units on a substantial scale (£12,715 in 1961-62). The cost is divided equally between the Health and Children's votes: it is perhaps symptomatic of some lack of co-ordination that Welfare does not contribute, as in the Hertfordshire scheme.

In *Essex* problem families in their own homes are dealt with by the health visitors supported by specially selected home helps, the charge for whose service is waived where considered desirable. Families are occasionally sent to voluntary rehabilitation centres (e.g. the Mayflower, Brentwood, Spofforth Hall, Crowley House and Dr. Barnardo's, Barkingside, all outside the London area except the last).

In the Welfare Department there is a family case worker responsible for undertaking the rehabilitation of homeless families at the Willows (which provides temporary accommodation for some 50 families). There is also a family case worker at Harlow New Town dealing with the problems of the families newly settled there.

A description of the work with problem families of the Families Welfare Officers in *Hertfordshire* is given in Chapter Six.

Kent, in addition to a Family Help Service, primarily intended to prevent the taking of children into care during the temporary absence of the mother, also has a Child Care Service providing "organised and constructive facilities for the rehabilitation of problem families". During 1960, 61 families commenced the initial service and 12 the follow up service. To quote the Medical Officer of Health's report "an assessment of the results shows that in 62 per cent of the cases in which the service started during the year the families were rehabilitated to such an extent as to take them out of the problem category". An occasional case is also sent to a voluntary home.

In *Middlesex*, three experimental schemes are in operation. In one area, two specialist health visitors work with problem families only. In a second, a Family Service Unit operates as agent for the Council. In a third, selected home helps, more highly paid than ordinary home helps, after a week's special training course, undertake either full time domestic work during the day, or if necessary, resident or full time non-resident help in a family where the mother is unable to manage her home and family or is away from home through illness. It is hoped to extend this last service throughout the county: an extension of the first scheme is inhibited by "the very great difficulty in recruiting suitable health visiting staff of the type prepared to undertake this type of work exclusively."

Surrey, like West Ham, set up in 1956 a committee on this question ("Working Party on the Prevention of Break up of Families"), but on a more elaborate scale, since it consisted of a number of elected members and of officers of both the county and of the country districts, with representations of a number of voluntary bodies, to a total of some 40, and with the Clerk of the County Council as secretary.

Following the recommendations of this body, a special service for problem families has been set up, based on a system of co-ordinating conferences, meeting not less frequently than quarterly, under the nine Divisional Medical Officers. Each Division (except one where the field work is done by a Family Service Unit) has a Divisional Social Worker, whose duties consist in visiting and dealing generally with problem families. An authorised establishment of up to the equivalent of 12 full time specially selected home helps is available to assist them. A training home on the lines of London "special units" has been established at Banstead, and use is also made of the voluntary homes.

It will be seen from the foregoing pages that questions of homelessness, child neglect and problem families are closely interrelated, that Public Health, Welfare and Children's

Departments all take a hand in trying to sort them out, and that the rôle of these departments and the degree of co-operation between them varies considerably in the different authorities. No Act of Parliament at present gives specific powers to deal with these problem families: Welfare Departments do so under Section 21 (1)(b) of the National Assistance Act (power to provide temporary accommodation); Public Health under Section 24 (provision of health visitors) 29 (of home helps) and 28 (prevention of illness and care or after care) of the National Health Service Act; Children's Departments under the general powers to consider taking into care in cases of neglect (Children's Act, 1948 1 (1)(b) and 41(5): see also Home Office Circular 160/'48).

The report of the Committee on Children and Young Persons 1960 (Ingleby Committee, Command 1191) however recommended that there should be a statutory obligation on counties and county boroughs to submit for ministerial approval schemes for the prevention of suffering of children in their own homes, to include:

(a) the detection of families where help is needed

(b) the co-ordination of information (investigation) and diagnosis of the problem.

(c) the provision of appropriate assistance

and (d) arrangements for making the services known to the public and for advice as to where individuals can apply for help.

The Home Secretary has indicated that preparations are being made to implement these recommendations, and legislation was adumbrated in the Queen's Speech in October 1962.

Before making its recommendations however, the Committee stressed the need to "recognise both the obviously inadequate or substandard family and the much less obvious family in which there is maladjustment of personal relationships: both the classical 'problem family' and what might be called the 'family with a problem'. In the former the standard of behaviour and morals will sometimes be as deplorable as the material conditions, but personal relationships may remain good and helpful to the children. In the latter, though outwardly seems all as it should be, the disturbance of family relationships may be a real danger to them".

Although the Committee "because of the varying conditions and requirements in different areas" did not advocate any uniform machinery for co-ordination of local authority services, there seems some implication that both these types of families are primarily the concern of Children's Departments. They do however require very different handling. The "classical" problem family of the sort defined at the beginning of this section, or rather the parents in it, are generally conscious of being in a mess, and if wisely handled, willing enough to try to profit by assistance to get out of the mess. Where the suffering and neglect which children experience is mental and not environmental, the reaction of the parents to outside interference is very different. In both cases there is an adult rather than a child problem.

The "classical" problem family might well be best handled by a Welfare Department constituted on the lines suggested in Chapter Seven. The other type would seem to present a medical and more particularly a psychological problem, more suitably tackled by a Public Health Department, perhaps through child guidance and marriage guidance clinics.

As has so often happened, the pioneers in a constructive approach towards problem families were a voluntary body, in this case Family Service Units, already mentioned as receiving support from the County Councils of Middlesex, London and Surrey, which are the only Metropolitan areas in which they at present operate. They were the first to recognise that what these families need above all is practical help and advice. For no other highly-skilled occupation may the new entrant be expected to start work without training and adequate tools, but in the typical problem family care a young woman with a number of children on a low income who may herself have been a member of a problem family, is expected to master a situation which a middle class housewife with a stable background finds sufficiently exacting.

IV. "DOWN AND OUTS"

Rogues and vagabonds, tramps, vagrants, casuals, down and outs—this group of the population has figured in legislation under a variety of names ever since the sixteenth century. Until the National Assistance Act 1948, they were the responsibility of local government—first of the parishes, then of the boards of guardians and finally, from 1930-1948, of the county and county borough councils. Now they are that of the National Assistance Board, their official title is "persons without a settled way of living", and they are no longer accommodated in casual wards but in "reception centres".

Except in the County of London they are an almost negligible problem, the only reception centres left in the home counties being in Hertfordshire, Kent (1 each) and Surrey (2), all well outside the Metropolitan area.

Inside the county the problem, once known as the "embankment problem" still exists, and ever since 1935 special machinery has existed for dealing with it. At that time the homeless numbered about 3,000 nightly, and the London County Council set up a body known as the Consultative Committee on the Homeless Poor to bring together all the various official and voluntary bodies dealing with them. This committee recommended the establishment of what was then known as the "Embankment Night Office" where applicants could go or be referred for help, and a register of them kept.

The Consultative Committee still functions, and this office, now known as the Welfare Office for the Homeless still exists under Charing Cross railway bridge, though it is now open during normal office hours, instead of all night as it was at one time. The expenses of its upkeep are shared between the L.C.C. (Welfare Department) and the National Assistance Board.

Table XV shows the number of applicants during the five weeks ending 31st March, 1962 and how they were dealt with.

Of the vouchers for board and lodging given, 218 were issued by the British Legion, 55 by St. Martin's in the Fields, 10 by the Royal Artillery Charitable Fund and three by the

TABLE XV.—Applicants at the Welfare Office for the Homeless, During Five Weeks to 31st March, 1962

Where sent	*Men*	*Young Men*	*Women*	*Young Women*	*Children*	*Total*
			Numbers sent			
Church Army	2	—	1	3	—	6
Salvation Army	148	1	1	—	—	150
Providence Row Night Refuge	4	—	5	—	—	9
S.O.S. Society	1	—	—	—	—	1
St. Barnabas-in-Soho	—	—	2	—	—	2
Whitechapel Mission	—	7	—	—	—	7
Common Lodging Houses	3	—	—	—	—	3
Camberwell Reception Centre	115	16	—	—	1	132
Southwark Reception Centre	—	—	22	7	—	29
Admitting Officers	10	—	12	4	26	52
Children's Officer	—	—	—	—	10	10
Given vouchers	278	2	4	2	—	286
	561	26	47	16	37	687

Royal Engineers Benevolent Fund, and of the recipients 257 proceeded to the Salvation Army, 21 to the Church Army, four to L.C.C. Lodging Houses, three to Common Lodging Houses, and one to St. Barnabas.

In addition to the 680 provided with accommodation, 203 were assisted with advice, of whom 50 were referred to the Assistance Board for immediate grants.

The average daily number of homeless applicants was about 20. A year previously the number had averaged only about eight, the figures fluctuating with a close relationship to the unemployment index.

It will be seen that while the majority of applicants were referred to voluntary organisations, 132 men and 29 women were sent to reception centres. These are administered by the London County Council as agents for the National Assistance Board. At both the men's and women's centres there are welfare officers to assist the residents to secure work, and at Camberwell there is an officer from the Ministry of Labour in daily attendance for the same purpose. On being found work, a man can go to his job daily from the centre, being given his breakfast, evening meal and all meals at week-ends, and on receiving his first week's wages, if unable to find lodgings, can remain in the hostel established at the centre at a low charge.

On 31st March, 1962, there were 351 men and seven women in the reception centres, and 37 men in the hostel. The turnover in the reception centres is large, admissions and discharges numbering about 8,000 a month.

There is a suggestion in the Press at the present time that sleeping out in the Metropolis is on the increase, more particularly in station waiting rooms, and that those concerned are of a new type, being younger people in search of work or actually in work, rather than old type of embankment "down and out". As such, they are unwilling to go to the reception centres frequented by and more or less acceptable to the latter. In so far as this increase is a fact, and the Consultative Committee is about to look into the situation, this would seem to be a further symptom of the acute housing shortage in London, and of the need for more hostels, commented on in Chapter Six.

It is clear that the Central Consultative Committee and the organisations associated with it, performs a valuable function which could not be allowed to lapse with the disappearance of the London County Council. An obvious solution would be for the Westminster group of boroughs to take over the Welfare Office and for the National Assistance Board to manage the reception centres direct.

The groups of the socially handicapped discussed in this chapter raises some interesting problems regarding the organisation of the welfare services. Welfare, Health and Children's Departments all concern themselves with problem families and with the unmarried mother and her child, and suggestions are now being made that Children's Departments should increasingly concern themselves with adults by taking preventive action in homes where children are regarded as being at risk. Hostels and common lodging houses differ from normal housing in providing residential accommodation for categories with special social needs rather than dwellings in which families and individuals can lead wholly independent lives.

The suggestions made in Chapter Seven and Table XIX envisage Welfare Departments responsible for all *adults* with special welfare needs, and responsible also for providing any special residential accommodation required by them, with Health Departments concerning themselves with those whose primary need is for medical and nursing attention, and Children's Departments with *children* deprived of a normal home life. But they also envisage the closest co-operation between Welfare, Health, Children's and Housing Departments, since their functions so frequently overlap.

CHAPTER SIX

Housing and the Welfare Services

I. HOUSING AND THE OLD

IT has already been said that whereas most of the welfare services are the responsibility of the first-tier authorities, housing, save in the case of the County Boroughs and the County of London, is the responsibility of the second-tier authorities. Indeed, housing, until comparatively recently, has hardly been held to be a welfare service at all. The production of houses has been regarded as the function of the borough engineer, the collection of rents that of the borough treasurer. Housing managers with a training in the social implications of a housing programme and housing management, are something of an innovation, and there are still many housing authorities without one. Even when the housing and welfare authority is one and the same, there may be little communication between the departments concerned. An enquirer into the facts about the homeless at the Housing Department at London's County Hall was told this was nothing to do with housing—he should address his questions to the Welfare Department.

This attitude has been due to the enormous demand for ordinary housing following the cessation of building during two world wars and the obsolescence of working class accommodation built in the last century. In the words of the Minister in 1958* "For years we have had to concentrate mainly on houses for the growing families." It is significant that it is in some of the rural areas, who have completed their normal housing programmes, that some of the most advanced provision for old people has been made, and significant again that in Denmark,

* Foreword to the Ministry of Housing and Local Government's "Flatlets for Old People," 1958.

where the housing problem is less acute, the old are the one category for whom municipal housing is provided.

It is only during the past decade that the housing of old people has received serious attention in this country. There was a general feeling immediately after 1948 that their needs had been taken care of by the National Assistance Act of that year, with its provision of residential accommodation for them, and the impression that this was to be of hotel standard, which it was presumed to be what old people themselves wanted. How far this is from being the case may be read in Peter Townsend's "The Last Refuge".

The census of 1951 revealed that of private households of one or two persons with the head of the household over the age of 60, 62·5 per cent occupied four rooms or more. This suggested that if more compact and suitable accommodation were provided for the old, many of these larger dwellings could be freed for families. At the same time social surveys up and down the country* indicated that the old, being economically the most vulnerable element in the population, were frequently being forced into the worst accommodation, and were being driven into local authority homes on this account, and not because they wished to relinquish their independence.

Thus increasing recognition came to be given to the importance of housing in the welfare of old people. The Ministry's Housing Manual of 1949 devoted one of its 100 pages of text to Dwellings for Aged Persons. Its "Housing for Special Purposes" in 1951 gave seven pages to the same subject and in 1958 the 20 page "Flatlets for Old People" was wholly devoted to it. Meanwhile a succession of circulars (including a joint one with the Ministry of Health emphasising the need for the coordination of housing and welfare provision) urged greater attention to this form of housing, and grant and more particularly subsidy policy was directed to favour it. "Flatlets for Old People" indicated that the proportion of one bedroom dwellings provided increased from 7·7 per cent in 1951 to 13·3 per

* At least 32 of these were made between 1945 and 1958 in all parts of great Britain. See, e.g., "Over Seventy," National Council of Social Service 1954. See also Peter Townsend, "The Last Refuge," p. 327.

TABLE XVI.—OLD PEOPLE'S HOUSING IN SOUTH-WEST ESSEX

Borough	1 *No. of bed sitting room dwellings*	2 *No. of 1 bedroom dwellings*	2 *% of 1 & 2 occupied by old people*	4 *Accommodation specially provided for old people* *Bed-sitter*	5 *1 bedroom*	6 *% of bed-sit. & 1 bedroom dwellings in ct. programme*	7 *No. of old people on waiting list*	8 *Remarks*
East Ham	757 (cols. 1 and 2 together)		33	20	—	25	350 (estimate)	
West Ham	52	923	42	8	48	40	no inf'n	
Barking	132 s'gle 74 double	870 flats 102 bngs.	80	16	78	20	222	Special housing all in group schemes, one with E.C.C. grant. (Another planned.)
Chigwell	30	182	70	23	77	in a high proportion	74	Group scheme with E.C.C. grant under consideration.
Chingford	21	117	bsrs. 70% 1 br dwlgs. 30%	—	—	—	97	Ditto
Dagenham	30	330	50–60%	30	117	14	415	
Hornchurch	75	63	96	75	63	26 (specif. for old people). 60	234	Old peoples' dwellings are in 9 groups each with a "contact". Scheme with E.C.C. grant under consideration.
Ilford	169	177	100	—	84* (see r'mks.)		319 (when list closed 2 years ago)	*Stoneleigh Ct. for 84 phys. handicapped mostly aged. (E.C.C. grant.)
Leyton	67	137	probably a minority	23	32	29	258	2 small schemes with E.C.C. grant.
Romford	263 (cols. 1 and 2 together)		73	10*	197	11	291	*The 10 bed sitters are in a scheme with E.C.C. grant. 4 more such schemes (62 beds) under construction
Walthamstow	280	1,171	90% bsrs. 75% 1 br.	112	139	25 (was 50)	312	Scheme with E.C.C. grant agreed.
Wanstead & Woodford	—	219	mostly	98	74	48	39 couples +189 indvls. over 55	Special provision includes 4 adapted houses run by W.V.S., and an adapted hotel (charge from 6gns. a week inclusive).

cent in 1957, and it is claimed that it has now reached 30 per cent (see however, pp. 102-3 on the relation of one bedroom dwellings to old people's accommodation).

Nevertheless, provision is still well short of demand, and varies greatly between authorities, while co-operation between Housing and Welfare Departments is generally limited.

To ascertain the position in all the nearly 100 housing authorities in the Greater London area was impossible within the scope of this study, but particulars were obtained from the 12 Essex authorities in the area, and are reproduced in Table XVI.

From this it will be seen that two boroughs only, Walthamstow and Romford, have more than 200 dwellings specially provided for the old (columns 4 and 5), though a third and much smaller one, Wanstead and Woodford, approaches this figure, as does Dagenham. The approximate number of persons over 65 in these boroughs in 1951 was 11,400, 8,200, 7,200 and 5,300. On the other hand, Leyton with 10,400 had only 55 dwellings specially for the old.

Ilford, with the biggest aged population of all (18,200) had only 84 and these are in fact dwellings erected for the physically handicapped, though a large majority of the inhabitants are aged.

In Ilford, however, all the Council's bed sitting room and one bedroom dwellings numbering 346 are occupied by old age pensioners (columns 1-3) while in Leyton, with a smaller number of dwellings (204) it was stated that probably only a minority of residents were aged.

On the other hand Walthamstow has provided 1,451 such dwellings, and Barking, though having a low proportion of old people, 1,188, and in both these boroughs about 80 per cent of the residents were said to be aged.

Conspicuously little provision for the old is made by West Ham, and this was agreed by the newly appointed housing manager there, who said however that in the current housing programme, 40 per cent of accommodation was in bed-sitting room and one bedroom flats (column 6). Only two other boroughs have higher figures than this—Ilford (60 per cent) and Wanstead (48 per cent). The percentages in Barking and Walthamstow are comparatively low which is not surprising in view of the extensive provision already made. Walthamstow in fact until quite recently was providing 50 per cent of this type of accommodation.

Wanstead and Woodford, which has a high proportion of old people in its population, appears to be the borough most alive to their needs. The present provision of one bedroom and bed-sitting accommodation is not extensive, though above average apart from Barking and Walthamstow, but the future programme of 48 per cent such dwellings is conspicuously large. Moreover, the provision of special dwellings for the old is notable not only in quantity but also in quality. It includes, among other accommodation provided by the borough itself, a converted hotel, where old people with some means are provided with bed-sitting rooms at an inclusive cost of 6 gns. a week upwards, under the supervision of a manageress and her assistants. A nursing home is about to be converted to a similar use. Four adapted houses are run by W.V.S. with the support of the local council, and there is also an Abbeyfield Society scheme in the borough grant-aided by the county council. It is also of interest that the waiting list of old people here is divided into couples of pensionable age and single (or widowed) persons over 55. With the departure of the younger members of the family the need for smaller accommodation for the ageing arises well before pensionable age, and this is not generally recognised. In this connection it is to be remarked that in Woodford they are normally able to rehouse homeless *families* in normal accommodation within six months, but *individuals* may have to wait for up to two years.

The least satisfactory position appears to be in Leyton, which has the highest proportion of old people in the area yet provides very little special accommodation for the aged. Nor are the majority of bed-sitting rooms and one bedroom dwellings occupied by the old. The 29 per cent of one bedroom and bed-sitting room dwellings in the current programme will take a long time to remedy the existing situation.

One point emerges clearly from this very limited survey, viz.

that the number of one bedroom dwellings is a highly unreliable index of the amount of provision made by an authority for its old people.

While they occupied all these dwellings in Ilford, in Leyton they occupied less than half, and in West Ham a special survey by the borough showed the proportion to be 42 per cent.

County councils, except London, are not housing authorities, but have the Minister's general consent (under Section 56 of the Local Government Act 1958) to contribute a maximum of £30 per house per annum towards the cost incurred by county district authorities in making special provision for old people. The following figures (from I.M.T.A. Welfare Statistics 1960-61) indicate the amount expended per 1,000 population by the five counties in the metropolitan area by way of grants to housing authorities for this purpose.

Essex	£3	6	0	Middlesex		Nil*	
Hertfordshire	£1	2	0	Surrey	£15	18	0
Kent	£3	2	0				

The average expenditure for English counties is £5 11s. od. the highest Dorset with £23 2s. od.

The London County Council's main contribution to special housing for old people has been in the form of colonies of bed-sitting room bungalows, numbering from 30-70, with quarters for a warden nearby, and a club room. There are 11 of these, all on out-county estates, except one in Putney, on the borders of the county. There are also three large dwellings converted into flatlets in co-operation with the W.V.S.

Of the 113,000 dwellings built since the war by the London County Council, 5,300 have been bed-sitting room accommodation for the old, and 13,000 one bedroom dwellings, about 60 per cent occupied by the old. The present programme comprises 5 per cent bed-sitting room dwellings for the aged, 10 per cent one bedroom dwellings for them, and 5 per cent one bedroom dwellings, for others.

The estates officers keep an eye on the old people when collecting their rents, and see that they have readily available the cards marked with a large H, with which they have all been

* Middlesex has since approved grant aid to six schemes.

supplied, and which they can put in their windows to summon help in an emergency.

There is however, a waiting list of 2,500 for one bedroom dwellings, 72 having been on it since before 1946. Cases remain long on the waiting list because the only available accommodation is frequently on out-county estates, or otherwise at a distance from their familiar environment. There is undoubtedly a need for more specialised accommodation for the aged within the county boundary, but the desiderata for dwellings for old people laid down by the Ministry in "Flatlets for Old People" can only be met by special and concentrated provision. This means some degree of segregation of the old, and against this there is a strong prejudice in this country. This is not so in Scandinavia. In Copenhagen for example, there are 7,000 old people on 12 housing estates, each containing 250 to 800 flats in up to 10 separate blocks, spread over the whole of the city so that the residents can maintain their local contacts. Since there is plenty of evidence that this maintenance of local contacts is one of the chief desires of old people, there seems a strong case for experimenting with at least one or two blocks of flats for the old in central London. If 70 old peoples' dwellings can be segregated on an out-county estate, why not 70 flatlets in a block in inner London, especially since recent research* has shown that old people on the whole are by no means averse from flats on upper storeys and the use of lifts.

The London County Council has experimented with one type of provision in the central area, by building two storey blocks of old peoples' flats in the curtilage of an old peoples' home in Hackney. There are conflicting views as to the success of this. Some of the residents appreciate the availability of a mid-day meal in the home and of nearby assistance in case of illness. But the association of the independent and the dependent on a fairly confined site is not altogether happy, and the idea that the flat residents would more readily enter a familiar home when they became infirm has not quite worked out.

Few old people in suitable accommodation wish to relinquish

* "Dwellings for Old People", Hole & Allen, Architect's Journal, 9th May, 1962.

their independence until they have to go to hospital with a terminal illness, and since the home is always full, a vacancy is unlikely to be available if ever wanted.

Croydon has at present 800 of its 11,000 dwellings specially provided for old people, and one bedroom and bed-sitting room flats comprise 20-25 per cent of its current programme. It has one two-storied block of 16 flats with a warden and club room, originally built for aged couples with the aid of a grant from the Lord Mayor's Fund, but now housing eight couples, six widows and two widowers. Further provision of this kind is under consideration. Six large dwelling houses have also been adapted as flatlets for old people, and there are two welfare officers in the Housing Department, both social science trained, one spending much of her time in looking after the old.

Voluntary and charitable effort has long made a substantial contribution towards the housing of old people, notably in the form of almshouses, of which there are a large number in the London area. Since the war, the Church Army has provided its Churchill Houses. A number of other housing societies have been formed, most of which earlier on concerned themselves with the provision of residential homes for the old, but some are now providing housing, for example Hearth and Home Ltd. and the Abbeyfield Society. Three schemes are receiving grants from the London County Council under the Local Government Act 1958. With the latest Housing Act encouraging the efforts of housing societies, it is likely that more provision of this kind will be made by them in the future, especially since this is the only form of housing in which they can receive subsidy.

II. HOUSING THE HANDICAPPED

While there has been an increasing volume of exhortation and advice from the Ministry on the subject of housing for the old, commendation of the special needs of the handicapped has been limited to a single page in "Housing for Special Purposes" 1951. This perhaps reflects the numerical importance of the two classes, but suitable housing for the handicapped is even more decisive in their happiness than it is for the old, since unsuitable housing in the case of a cripple may mean virtual imprisonment. Normal people can put up with the inconvenience of conventional dwellings, with their proliferation of stairs, inconveniently sited bathrooms and lavatories, and ill-planned kitchens, but for the handicapped they may make life nearly intolerable.

More attention is now being given to this question, at least in unofficial circles. A "Research House" has been privately presented to the Nuffield Orthopaedic Centre at Oxford intended for the study of the whole subject of housing, appliances and furniture design in relation to different types of disability, while a Charitable Trust has undertaken to provide a capital sum and an annual income for five years "for the purpose of experimenting in the design of dwellings, for the use of those with impaired function and for the development of components for such dwellings, and of domestic appliances for special requirements for such dwellings."

Since the outcome of these projects may well be the emergence of a model scheme which will eliminate planning errors in a field which has been little explored in this country, the delay in provision may have its advantages.

Meanwhile it appears that the only authority in the London area which has provided accommodation specifically for the disabled (apart from the Ilford scheme mentioned above) is Croydon. This it has done in a number of specially adapted units on its housing estates, and also in a group of six purpose built bungalows, each containing a sitting room, domestic offices, 2-3 bedrooms and a roomy garage with an interior as well as an exterior door.

The rents here are 50/- plus rates, subject to a rebate to one fifth of income. Tenants are received on medical recommendation and there has never been a lack of them. The children on the estate where the bungalows are located exhibited a certain amount of curiosity initially, but this soon passed.

Reluctance to provide this special accommodation is sometimes due to the problem which arises when the cripple dies or moves. It would seem this might be simply overcome if it were made clear at the initial letting that tenancy was subject to occupation by someone able to profit by the special amenities, and when this no longer obtained alternative accommodation would be provided.

The remaining authorities in the London area rely on the provision of ground floor accommodation, and adaptations of existing premises (by the provision of ramps, wide doors and special fittings) to meet the needs of the handicapped, except that the London County Council is providing a special hostel for six adjoining its workshop for the handicapped.

In the voluntary field, the St. Giles Housing Society has just opened six bed-sitting room flatlets, specially planned for the handicapped in an adapted house in Clapham, on two storeys with a lift, and with garage space adjoining.

III. HOSTELS AND LODGING HOUSES

(a) *Hostels*

"Housing for Special Purposes" can almost be regarded as a handbook on the relations between housing and welfare. Two of the chapters in it, on housing the aged and the handicapped, are commented on above. Two others, "Housing Needs of District Nurses and Midwives" (mainly in rural districts) and "Houses for Higher Income Groups", have clearly little or nothing to do with welfare, and in any case are dealt with very summarily. The remaining chapter however, on "Housing Needs of Single Workers, Students and Apprentices" again has a welfare angle. The Housing Manual Sub-Committee which produced the handbook stated that it had "a very clear idea of the sort of accommodation, and its defects, which the single person of limited means has too often to be content with—the comfortless and solitary room in an old fashioned lodging house, never designed for that purpose, and frequently not conducted with any understanding of or interest in the occupant's comfort. . . . We consider that in general the accommodation suited to the great majority of single persons who do not live with their parents or other relations will be hostel accommodation."

The key words in the foregoing are "the single person of

limited means". Obviously included in this category are students and apprentices. Another less obvious group are the married provincials who have to come up to London on a job or on a course, and have for a time to maintain themselves and their families separately.

Hostel accommodation for these classes is provided by the London County Council, though not by any other authority, and although now provided under the Housing Acts, is, significantly, administered by the Welfare Department. Its origin is of interest. All the five men's hostels were originally casual wards which fell into disuse after the war owing to social changes. They afforded good kitchen and dining room quarters, excellent bathing facilities and reasonable sitting rooms, but the bedroom accommodation, consisting of the old cells measuring about 8 ft. by 5 ft. was cramped. Nevertheless it was felt that all five, with a total of 425 beds, would offer acceptable accommodation to an entirely different class after adaptation. In fact they are always full. Three women's hostels, with 233 beds are also provided, though not in old casual wards, and these are also always full. Present charges are 66/6d. and 70/- a week for men, including breakfast and supper and all week-end meals, and 66/6d. to 77/- for women, according to whether sleeping is in single rooms or dormitories. In the last financial year, income of the hostels was £111,508 against expenditure of £130,717. Hostels for younger men and women are provided by a variety of voluntary organisations, such as the Y.M.C.A. and the Y.W.C.A. and Toc H. One of the great housing trusts has also recently built one intended more especially for older women.

(b) *Lodging houses*

Municipal lodging houses are maintained only by London, with three accommodating 1,875, and Croydon, with one for 100. The charges in London are 3/- a night, 19/- a week at one, 2/9d. a night, 17/- a week at the others, with a few special cubicles at a rather higher charge in two of them. In Croydon the charge is 2/6 a night, 16/6 a week, but 2/- and 8/- a week for old age pensioners. In London, income was £131,219 a year against expenditure £160,345, in Croydon £3,665 against

£5,527. In London, the lodging houses are administered by the Welfare Department, in Croydon by the Public Health Dept.

All of them are practically always full. It is estimated that some 80 per cent of the residents are permanent, and these comprise all sorts and conditions of men, from those in regular employment, including white collar workers who choose this way of life, to old age pensioners, who find it the cheapest way of maintaining their independence.

A few of the regulars, and many of the casuals, who also frequent the charitably and privately maintained common lodging homes, are social casualties, from prisons, mental hospitals and broken homes. A recent study* of these establishments suggest that they are not organised to give such men the help they need. Following this study, the City Parochial Foundation has set aside £50,000 for the building of a new type of lodging house for this type of case.

There are no municipal lodging houses for women. The London County Council has considered their provision, but concluded they are not needed at present on the grounds that there is an adequate supply of this type of accommodation, mainly through the voluntary organisation "Cecil Houses".

The demand certainly exists in the London area for more hostels for both men and women, and for more lodging houses for men, perhaps for women too. Since sites in London are always in short supply, and since hostels are one of the most compact forms of housing that can be built, it would seem that there is scope for a good deal more of this type of provision by local authorities in the London area.

IV. THE HOMELESS

Section 21(b) of the National Assistance Act 1948, requires counties and county boroughs to provide "temporary accommodation for persons who are in urgent need thereof, being need arising in circumstances which could not reasonably have been foreseen, or in such other circumstances as the authority may in any particular case determine." Ministry of Health

* Merfyn Turner, "Forgotten Men," National Council of Social Service.

Circular 87/48 issued to local authorities in explanation of the Act, made it clear in para. 15 that this accommodation was primarily intended to deal with the victims of disaster such as flood and fire. Eviction was also mentioned, but it was specifically stated that this provision was not one for dealing with the inadequately housed.

Nevertheless, the main use to which temporary accommodation provided under the Act has been put by all authorities has been to house the homeless.

What has occurred in Essex may be taken as an example of the outcome of this situation.

In its written evidence to the Royal Commission the Essex County Council "strongly" suggested that dealing with the homeless (apart from those resulting from major disasters, e.g. flood and fire) should be a function to be directly conferred on the housing authorities, and should not be the responsibility of the county council.

In the course of the oral evidence, a good deal of discussion took place between representatives of the Council and members of the Commission as to where these responsibilities really did lie. County representatives pointed out that the Act only gave them responsibility for those who were homeless through unforeseen circumstances, and eviction through non-payment of rent or other causes which occasioned the greater part of the homelessness, could not be said to be unforeseen. Commission members doubted whether the housing authorities had a legal duty to deal with such homelessness, and the final word on the subject was that "it may be in precise terms nobody's duty at the present time".

In the circumstances it is not surprising that there was a long succession of discussions and conferences between the county and the housing authorities in Essex on this subject. The outcome of these had been a gradual recognition by the housing authorities that homelessness, other than that caused by disaster, is properly their responsibility. At the present time, most of them will accept homeless families from the county on a guarantee of two years rent, and many of the boroughs in south west Essex deal with their own homeless problem without

TABLE XVII.—PROVISION FOR THE HOMELESS IN SOUTH WEST ESSEX

Boroughs	*No. of half way houses*	*Accommodation (families)*	*Housing Wait*	*Housing Rule* Length of residence in borough required for re-housing	*Notes*
East Ham	2 blocks in an old hospital	25	18 months	1 year	
West Ham	2	8 families	7 months	10 years	
Barking	No. special accommodation	—	usually at once	2 years	Use vacant acquired property.
Chigwell	None	See note	about 9 months	4 years	Use odd bits of property.
Chingford	None	—	1–2 years	3 months	
Dagenham	1	9	1–2 years	none, but do not look favourably on less than 10 years.	
Hornchurch	None	—	—	12 months	
Ilford	2	18	9 months	12 months and then on waiting list at least for 1 year.	
Leyton	None	See note	12 months	None for permanent residents of borough.	Use odd bits of property for evicted.
Romford	None	—	—	Applications received from any borough residents in actual housing need.	Relies on provision by Essex County Council.
Walthamstow	3 half way houses 55 acquired houses	15 55	12 months	3 years normally (but any application accepted).	
Wanstead & Woodford	2	12	6 months (families) longer for individs. or couples, up to 2 years (Due to shortage of small accommodation).	7 years	

references to the county, by placing them in their own half-way houses* or in sub-standard accommodation.

As a result, during the year 1960-61, twice as many families went out of the county's Part III accommodation as in any previous year, and the situation in which families might remain in such accommodation for seven or eight years has now passed.

Walthamstow, where the housing manager is secretary of the local group of housing managers, has been the pioneer in this movement, and there the borough absorbs virtually all its own homeless: indeed, it has been able to help out the County Welfare Department by taking certain cases from Part III including one a year of those families landing at a port in the county without any housing claim.

Table XVII, overleaf indicates the present arrangements for the homeless in the various boroughs. Most still have half-way houses, but there is a general tendency to close these down and admit direct to substandard accommodation. It will be observed that nearly all boroughs reckon to have their homeless families in permanent accommodation within a year.

A factor which enters into the housing both of old people and of the homeless, is the length of residence in the borough for acceptance on the housing waiting list. It will be seen that this varies very widely.

This brief survey of developments in Essex indicates that homelessness is primarily a housing question. This is confirmed by the figures in the following table:

TABLE XVIII.—AVAILABILITY OF PRIVATE DWELLINGS AND HOMELESSNESS IN THE GREATER LONDON AREA

Local authority	*No. of private dwellings per 100 private households*		*No. of persons in temporary (Part III) accomm. per 100,000 population*	
	1951	1961	31.12.60	*early* 1962
Kent	95	102	5	10
Hertfordshire	95	101	23	23
Essex	92	100	16	16
Surrey	93	100	11	13
Croydon	87	96	11	18
East Ham	83	94	21	16
Middlesex	85	93	29	29
West Ham	76	90	26	47
London	71	85	71	108

* "Half-way houses" are establishments where, pending permanent re-housing families (including husbands) can live together and do their own house-keeping, but with limited accommodation and facilities.

The figures in the first two columns are derived from the Preliminary Report of the 1961 census, and show first that everywhere in the Greater London area the ratio of dwellings to households substantially increased in the decade, second, that in four authority areas the number of dwellings now equals or exceeds the number of households. Columns 3 (taken from I.M.T.A. Welfare Service Statistics 1960-61) and 4 (from L.C.C. Agenda papers, p. 39-40, 2nd July, 1962) show that it is with the authorities where the ratio of dwellings to households is lowest that the homeless problem is most acute. One factor here is the insecurity of tenants in the large Victorian houses divided into tenements.

Hertfordshire stands out as an exception, and the explanation here may lie in its different method of tackling the homeless problem, which is touched on below.

While basically practically all homelessness is a housing problem, since the only satisfactory eventual solution is the provision of a home, welfare considerations also enter in. The causes of homelessness graduate from sheer misfortune through irresponsibility to intolerably anti-social conduct. In other words, among the homeless are included the so-called problem families. Their proportion is a matter of dispute, but it is certain that the greater the housing problem the higher will be the proportion of "ordinary decent folk" among the homeless, and conversely, where the homeless are few, the greater the likelihood a considerable proportion will require some form of rehabilitation before they can be expected to retain any accommodation found for them.

In the early years after 1948, nearly all authorities provided their temporary accommodation in the old institutions, and because these afforded little or no privacy, only the women and children under 16 of applicant families were admitted, husbands and older children being required to find their own accommodation, which they could normally be expected to do without much difficulty. This, though never overtly proclaimed as a method of deterrence, undoubtedly acted as one, but more effectively in respect of united families than others, and therefore with socially undesirable effects.

Most authorities now have specially adapted accommodation, and some, but not all, admit husbands with their families.

As might be expected, it is in the county boroughs that co-operation between Welfare and Housing Departments is closest.

In *West Ham*, following a joint report to the Council on "unsatisfactory tenants and problem families" by the Town Clerk, Borough Treasurer, Medical Officer of Health, Housing Manager, Chief Welfare Officer and Children's Officer, the Housing Department has unreservedly accepted the responsibility of housing the homeless. In particular, the chief officers recommended that accommodation in which to attempt rehabilitation work should be provided by the Housing Department. Temporary accommodation is maintained by the Welfare Department, hitherto in the curtilage of former institutions, but now to be in separate accommodation. Husbands are admitted. Families are transferred from this to two half-way houses (accommodating eight families) in the housing department, thence to normal housing in due course.

In *East Ham* four small houses are used as temporary accommodation, with an occasional overflow into West Ham accommodation, and husbands are accepted. Stay is strictly limited to three months, by which time the family has normally been accepted in the Housing Department's half-way accommodation. This houses 25 families, the type of provision differing little from that in the Welfare Department, the distinction being that the cases received are definitely accepted for rehousing, which normally takes place within 18 months.

Croydon has stringent rules for the admission of the homeless. No husbands are admitted: no "blameworthy" cases (homelessness due to quarrels, rent arrears, etc.), unless with children under five; stay strictly limited to six months. These conditions were introduced at a time when there were some 60 homeless families accommodated, and the position was felt to be getting out of hand. Initially this policy incurred a good deal of local hostility which has now abated, since it was accompanied by some comprehensive welfare provisions. There is now one "Receiving Home" with accommodation for 14 families, who are under the supervision of a specialist health visitor (seconded from the Health Department but under the direction of the Chief Welfare Officer), whose job it is to instruct the women in budgeting, housekeeping and child care. Since the Housing Department will not accept cases with rent arrears, she also induces them to pay these off. In addition to dealing with those in the receiving home, she also visits and advises potential homeless families referred from the various Council Departments and elsewhere. Liaison with the Housing Department is very close. Wherever possible, the two chief officers co-operate to prevent admission to the receiving home, but as soon as this takes place, Housing is informed and the case is dealt with at the next fortnightly meeting of the Housing Sub-committee (a virtual impossibility where temporary accommodation is the responsibility of one authority and housing of another).

The Housing Department is proposing to close down its half-way houses, of which it had four, and deal with the homeless by ordinary housing.

As has been said above, the county welfare authorities have a much more difficult task in disposing of the homeless from temporary accommodation than the county boroughs. The position in Essex has already been described. *Hertfordshire* has evolved a distinctive method of dealing with the problem. Applications for accommodation are in the first instance received by the divisional welfare officers. These, if they see no alternative to admission, refer the case to the Families Welfare Officer, who, with three or four assistants is on the staff of the Clerk of the Council, their salaries being defrayed equally by the Welfare, Health and Children's Departments. If she cannot find a solution, the family, without the husband, is admitted to one of two family units, accommodating 25 and 10 families respectively. Each unit is in charge of a warden and his wife, who undertake rehabilitation, and themselves in a number of cases find accommodation. Where they are unable to do this, the families, after having been brought up to reasonable social standards, are admitted, with their husbands, to one of 24 properties acquired by the Council—including 12 by loan from Development Corporations and four from other departments of the Council, some of which accommodate more than one family. From these they may in some cases be rehoused by the local housing authorities, but district councils are reluctant to take cases who have come into the county with no housing claim and the scheme is always under threat of a blockage for this reason. The families in all this accommodation are returned as homeless, but at the end of 1961 out of a total of about 190 only 50 were in the two family units.

In addition to their work on behalf of the families in temporary accommodation, the Families' Welfare Officers are constantly approached by local housing authorities to investigate cases of rent arrears where eviction is likely. Some 600 cases of this kind had been dealt with in the three years up to the end of 1961.

Surrey until recently admitted women and children, without the husbands, to institutional accommodation, but limited provision has now been made to permit the reception of complete families if suitable, and this it is proposed to extend. A special unit for problem families, on the lines of those in London, has been established. The Council has also agreed to provide the district councils with financial assistance to the extent of one third of their net loss in the provision of half-way accommodation, and this has met with an encouraging response.

As is indicated in the Royal Commission's Report, the relations of *Middlesex* with the housing authorities in the county are less happy. The effect on the homeless problem may be seen from the following extract from the county's own evidence (p. 140):

> "Many families evicted from their homes remain in the county council's temporary accommodation for long periods because they are unable to obtain permanent housing. For example, it was reported to the county council's Welfare Committee on 28th January, 1958, that there were 40 families who at that date had been in temporary accommodation provided by the county council for over two years (some of these for over six years).
>
> The demand for temporary accommodation is consistently greater than the places available and the position is aggra-

vated by the small turn-over of cases in the county council's hostels. In a proportion of cases the families need some form of rehabilitation before they are suitable for permanent housing but the great majority remain in the hostels simply because they cannot obtain a house.

The county council is not a housing authority and cannot, therefore, provide the necessary permanent accommodation and whilst 15 of the 26 district councils in the administrative county have each undertaken (subject to various conditions) to provide housing for one homeless family in temporary accommodation per year the district councils are not particularly concerned to alleviate the position and somewhat naturally prefer to deal with the problem of their own housing lists.

The present difficulties are accentuated by the fact that among the 26 housing authorities from whom permanent accommodation is sought by or on behalf of the families in temporary accommodation there are considerable differences in the schemes of priorities for the allocation of housing which have been adopted. Where rehabilitation is necessary it is of little use to attempt it if at the end there is no reasonable likelihood of a permanent home being available."

When its evidence was given, the county had eight special hostels for the homeless, with accommodation for 112 families, two for women and children only, the other six for husbands as well. The number has now been increased to 10, with accommodation for 130 families.

As is shown in Table XVIII *Kent* has a higher proportion of dwellings to households than any other authority in the London area, and a lower proportion of homeless. A hint of an additional reason why the county is not greatly troubled by the homeless is contained in this extract from its evidence to the Royal Commission:

"The amount of accommodation needed can best be provided by one central hostel. This is situated near Maidstone and provides 47 units of family accommodation which is rarely completely full. Temporary accommodation is only offered to women and children, since experience has shown that when husbands were taken in, they usually gave up work in the area in which they were living and rehabilitation then became extremely difficult.

"Experience also suggests that the provision of only one central hostel is in itself a spur to the families to find accommodation for themselves."

In view of the full study recently made of homelessness in the County of London, and published in the Council's Agenda paper of 2nd July, 1962, no attempt is made to deal with the situation in the county here. It should be mentioned however, that the research team's main recommendation was that to the extent that homelessness is a housing problem, its alleviation or solution should be the Housing Committee's responsibility, and that homeless families should remain in Welfare Department accommodation not more than a month.

The purport of this chapter has been to show how closely Housing is related to Welfare, the difficulties that are bound to arise when these are the functions of two different authorities, and the advantages that could accrue from their both becoming the responsibility of the new London Boroughs.

PART THREE

CHAPTER SEVEN

The Reorganisation of the Services in the New London Boroughs

I. A POSSIBLE PATTERN

THE foregoing chapters (two to six) attempt a picture of how the welfare services are administered in Greater London today. The effect on the mind of the detached observer may well be rather like that produced by prehistoric cave paintings. They are remarkable overall, individual achievements are outstanding, but the general impression is of confusion, with breaks here and overpaintings there.

Like the prehistoric paintings, the welfare services have accumulated over a period of time with no one setting a conscious pattern. One bit of legislation overlies another, and shadows of the old work show through. Different authorities put their individual stamp on different parts of the background.

This confused picture presented by the welfare services in action is the result of two main factors.

First, legislation over a period of years has been directed towards the relief of particular categories of the needy with which the public conscience has especially concerned itself at the time. Examples are the blind in 1920, deprived children in 1948, the mentally affected in 1959.

Secondly, the various Acts of Parliament have allocated responsibility for the various categories of the needy to different authorities. Central government, as already stated, is responsible for financial assistance to all classes, and for the hospital and general practitioner services. Most of the personal welfare services are placed on the first-tier authorities, but there are exceptions. And while the Children's Act obliges each authority to have a Children's Committee and a Children's Officer, the National Assistance Act, 1948 (Third Schedule), permits authorities to discharge their functions under the Act either by an *ad hoc* committee, or with the Minister's consent, to some other committee. All the county councils and county boroughs in the London area have welfare committees except Kent, where the functions are referred to the Health Committee and there is no Chief Welfare Officer. The pattern in the country generally varies greatly (see Report of the Working Party on Social Workers (Younghusband Report), Table 38), though most authorities have both a Welfare Committee and a Chief Welfare Officer. All have Medical Officers of Health, but so also do the lower tier authorities—the boroughs and urban and rural district councils.

The welfare services provided by these lower tier authorities are housing and certain forms of assistance to the aged, under Section 31 of the National Assistance Act, 1948, and under Public Health Acts. The trouble with this sort of organisation is that human beings, even when in need, do not always fit, or remain, in the categories to which they are allotted. Physical disablement may be accompanied by mental disablement, and both may be the fundamental cause of social disablement.* All types of disabled may become old, while on the other hand the great majority of the old, (taking the criterion as pensionable age) are able to take care of themselves. All people require housing, and some of the disabled and the old, particular kinds of housing.

Secondly, while the responsibilities imposed by some Acts on the local authorities are clear and specific, others are vague and general. All authorities, central and local, and nearly all departments, have their welfare officers, from almoners in the national hospitals to housing welfare officers in the boroughs and county districts, and in these circumstances it

* A survey recently carried out by the Welfare Department of the London County Council indicated that about 30 per cent of the registered blind are physically handicapped in other ways; of those registered blind in a given year about 67 per cent had gone blind after reaching the age of 70; while the physically handicapped include substantial numbers of those handicapped by the infirmities of age.

is not surprising that there is confusion, overlapping and gaps.

Perhaps the most interesting revelation is the extent to which functions carried out by one department in one area are performed by another department, or even another authority, in a different area.

In Surrey, for example, initial applications for the reception of children into care are taken, not by Child Care Officers, but by the district officers in the Welfare Department. Because a high proportion of applications are for temporary care only, during the short-term absence of the mother, it is considered that this saves the time of a trained child worker who, however, is called in whenever circumstances demand it.

In Surrey again, education welfare officers and not child care officers attend the juvenile courts to supply the background of family history.

In both Kent and Hertfordshire the district welfare officers act as mental welfare officers (formerly duly authorised officers).

In Essex, problem families are the responsibility of specialist officers in the Welfare Department. In Kent they are looked after by a special branch of the home help service. In London, Health, Welfare and Children's Departments all have specialist officers supervising them—the Welfare Department in certain special residential establishments, Children's and Health in the families' own homes. In Croydon a health visitor with a special aptitude for dealing with these families was seconded to the Welfare Department for the purpose.

Homeless families throughout the London area are dealt with by Welfare Departments, with a greater or lesser degree of co-operation from the housing authorities. But in Essex there is growing recognition that homelessness which is not the result of disaster such as fire or flood is really a matter for the housing authority.

In Leyton, the old get special attention from the health visitors, in Ilford, from an officer in the Public Health Department, in Dagenham, from one of the Clerk's Department, in Hertfordshire and Middlesex, from a special type of home help.

In Croydon, the Health and Welfare Committees have jointly approved plans for the construction of a workshop to provide occupation for both the physically and the mentally handicapped, to be under the direction of the Welfare Department.

All these examples illustrate the fact that the welfare services are essentially homogeneous. The fundamental needs of all human beings, whether handicapped or not, are the same—adequate shelter, food, warmth, companionship, and self-respect, which is in essence a consciousness of having some value in the world. It is the function of the welfare worker to try to see that these needs are met for the handicapped, though the different types of handicapped have special needs of their own, to meet which special skills and experience are required.

The piecemeal provisions of the welfare services, and their operation by a variety of authorities has obscured these basic realities. As a result, specialist skills have been wasted. A notable example is that of home teachers of the blind. A high proportion of the blind are old people whose chief need is social support and most of whom are unable to profit from special instruction. Yet ever since Circular 32 of 1951, the Ministry of Health has banned the use of blind home teachers for any but the blind, for fear of a dilution of the quality of service, which means that a large part of their time is occupied in social visitation.

Meals on wheels are ordinarily regarded as a service for old people and in a recent survey* of their provision 96·6 per cent of recipients were found to be over 60. Yet this is a service equally applicable to the physically handicapped, and it is difficult to believe a much higher proportion of younger people could not benefit from it.

Examples could be multiplied of services and facilities which are available to one category of the needy in an area, but not to other categories equally able to profit by them in the same area, while in an adjoining area a similar service is provided for another category. Thus a narrow view of the scope of the individual services has led to an uneconomic use both of personnel and of the facilities provided.

* Amelia Harris. Meals on Wheels for Old People. National Corporation for the Care of Old People.

The best prospect of improving these services, at present so largely unco-ordinated, patchy and extravagant in their use of resources, would be to integrate as many of them as possible under one local authority, and to implement the recommendations of the Younghusband Report on the use and training of social workers.

The constitution of the new London Boroughs affords a unique opportunity of achieving these objectives. The great majority of them will have more than one Town Hall at their disposal. It is to be hoped that one in each borough, preferably the most central, would house all the Welfare Departments, Welfare, Health, Children's and Housing. One of the greatest advantages of the new set-up will be the integration of Housing with the welfare services under one authority. Stress has already been laid on the close relationship which exists between these services, and their association under one roof should certainly contribute to greater efficiency. Similarly the performance by the new London Boroughs of functions in relation to the old, now split between the county councils on the one hand and the borough and district councils, in co-operation with voluntary organisations, on the other, should improve the service to this ever-growing section of the population, more especially if the suggested amendment to Section 29 of the National Assistance Act were effected. (See pp. 36–7).

One would envisage that at this central office would be located all the specialist welfare staff—the psychiatric social workers, home teachers of the blind, occupational therapists, child care officers, home help organisers, family case workers, public health inspectors, etc., all in close contact with one another and with the centrally administered services, more especially the hospitals.

In outlying districts offices would be located the general purpose social workers,* whose function in the welfare field would be not dissimilar from that of the general practitioner in

* The term "general purpose social worker" is used in the sense of the definition given in the last sentence of §655 of the Younghusband Report —"a trained worker who is employed to undertake a range of duties in the health and welfare services." See also p. 9 of the report.

TABLE XIX.—POSSIBLE SCHEME OF ORGANISATION OF A WELFARE DEPARTMENT IN A NEW LONDON BOROUGH

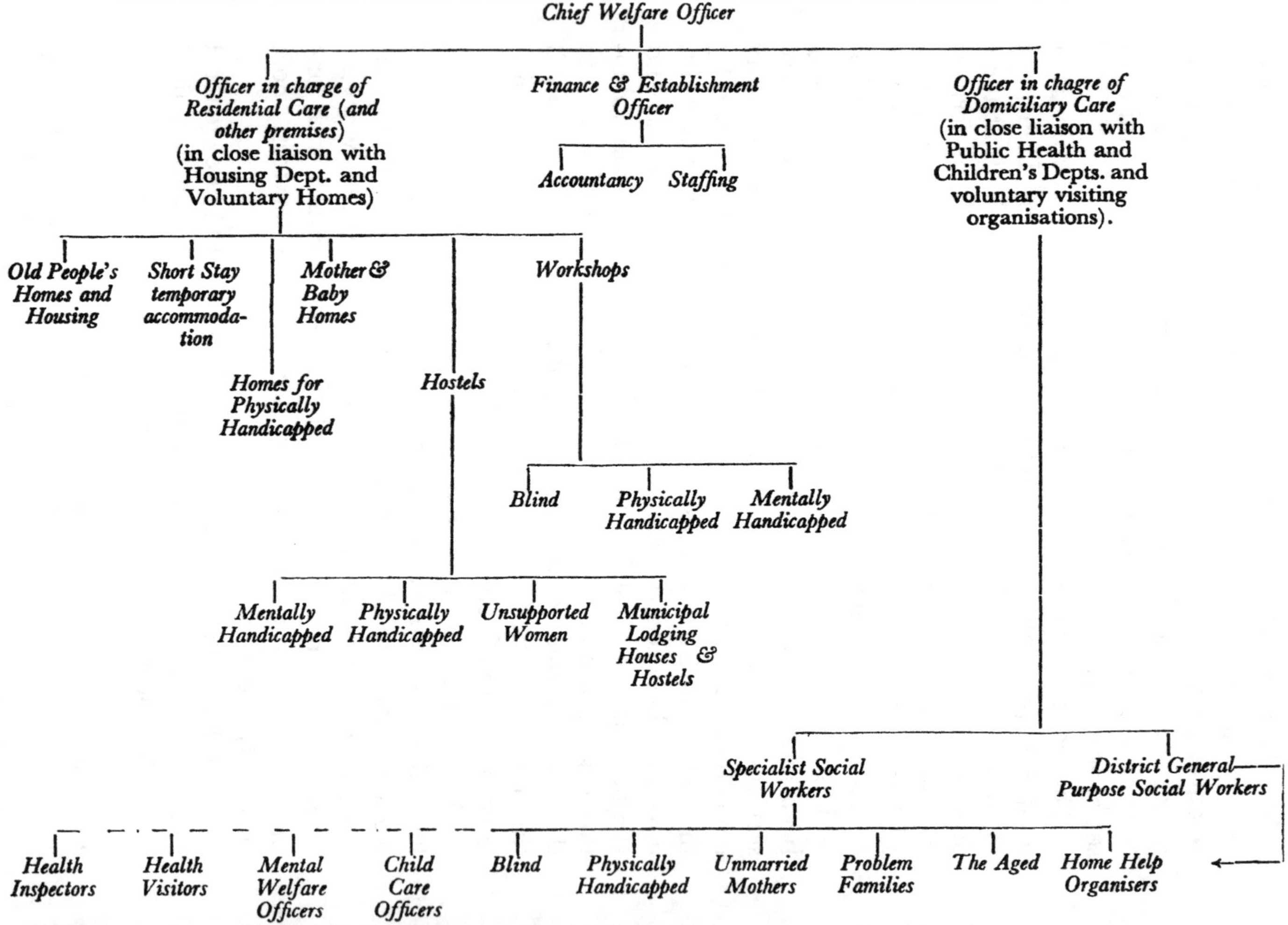

the medical field. He or she would receive the initial application, visit the home where indicated, make the diagnosis, ascertaining *all* the services required, and then continue to deal with the case himself, or, as the occasion warranted, call in the specialist who might then take it over, or prescribe a course of action for him to carry out. Behind him would be a range of residential accommodation available as hospitals are available to the doctors. Such general purpose social workers would normally be responsible for all initial applications and, subject to the intervention, when necessary, of the specialist workers, would be responsible with the aid of one or more welfare assistants for continuing oversight of old and physically handicapped recognised to be in need of help and of the stabilised cases of mental illness and mental retardation, and of social disability. He might well arrange short stay care in conjunction with the Children's Officer on the precedent set by Surrey. He would maintain close contact with the health visitors, home nurses and home helps in his district.

The number of such district officers required would, of course, vary with the size and nature of the borough, but Croydon offers an indication of the sort of provision needed. The work now done there by the welfare officers resembles more closely that likely to be required of the general purpose social workers than it does anywhere else, and in Croydon, with a population of 250,000, there are four districts. A suggested scheme for the organisation of a Welfare Department on these lines is given in Table XIX. On its relationship with other departments providing welfare services, see pp. 96-7.

A question which has received consideration for some time has been who should administer the welfare services. The Herbert Commission, following the Guillebaud Committee and the known views of the Ministry of Health, suggests they should be under the direction of the Medical Officer of Health, and reiterates the conception of a domiciliary team with the G.P. as leader. Those working in the field generally regard this as something of a chimera. General practitioners considerably outnumber health visitors and other local health and welfare workers, and this would make local sessions excessively difficult to arrange, even if the G.P.s had the time or inclination to attend them.

It seems likely however, that the welfare services will, at least initially, be placed under the control of Medical Officers of Health, for the simple reason that they are already in the field in the boroughs, and welfare officers are not. It will be a much simpler proposition to add welfare functions to whichever of the existing M.O.H.s is appointed to the post in a new Greater London Borough than to start a new department with a newly appointed officer. But although this is likely to be the initial arrangement, there are grounds for doubt as to whether it will or should be permanent.

Administratively, there is at least as strong a case for putting Children's Departments under Medical Officers of Health. On the one hand they are smaller and more homogeneous than Welfare Departments, on the other hand there is a growing emphasis today on the part preventive medicine has to play in the field of mental health by the easing of emotional tensions in childhood. That there is now no question of Children's Departments being administered by Medical Officers of Health is due to the Curtis Report* whose recommendation of the appointment of *ad hoc* Children's Committees and Children's Officers was accepted in the Children's Act 1948.

The wording of the Committee's recommendation is of interest: "We favour the establishment of an *ad hoc* committee reporting directly to the council (§ 440) . . . Our preference for the single *ad hoc* committee . . . is based in part on our desire that it should have its own executive officer with the standing of an important administrative official of the council. (§ 441) . . . The Children's Officer should in our opinion be highly qualified academically, if possible a graduate who has also a social science diploma . . . She should have marked administrative capacity and be able readily to grasp local government procedure and to work easily with local authority Committees (§ 446).

In other words the Curtis Committee was convinced that if local authorities were to make a success of their service for

* Report of the Care of Children Committee 1946, Cmd. 6922.

deprived children, its administration must be allocated to an independent committee served by a highly qualified and experienced chief officer. If this is true of the Children's Service, it is difficult to see why it is not true of the other local authority welfare services.

That there is official support for the administration of Welfare Departments by Medical Officers of Health, apart from the idea, which many believe to be illusory, that this would facilitate the emergence of the domiciliary team, seems to be due to two factors. First, the Ministry of Health in Whitehall administers both the National Health Service and the National Assistance Act as it affects the welfare services, and it would seem a natural corollary to a department in which the medical profession is strongly represented, that a similar set-up should function at the Town Hall. Second, there is a general impression that medical officers of health, by reason of their training in a liberal profession, tend to have a wider outlook than chief welfare officers, who are liable to be inbred in the traditions of the old Poor Law (which is not to say there are not myopic M.O.H.s and far seeing welfare officers). The Younghusband Report gives some confirmation to this notion, showing that of 177 welfare officers responsible for the administration of the welfare services with local authorities, 78 had no qualification except long service, and the possession of the relieving officers' certificate or the Scottish Poor Law Diploma was the qualification of 61 out of the 99 who had qualifications, while only seven had social science degrees, diplomas or certificates. The gradual implementation of the recommendations of the Report may, however, be expected to effect a marked change in this situation, while welfare officers would contend that an administrative flair and an understanding of human nature are more important than an academic training, though they may be enhanced by one.

Not all Medical Officers of Health are in favour of taking over welfare functions. Some consider that the existing scope of their departments is sufficient to demand their full attention, more especially in view of new calls made upon them under the Mental Health Act, 1959. Those favouring the take-over generally consider it essential they should have under them an officer of equivalent status to chief public health inspector to undertake the organisation of the welfare side of the work, including, of course, the maintenance of homes. The crucial question as to whether such an officer should have direct access to the committee gives rise to differences of opinion, as indeed does the question whether there should be a welfare committee, or whether it should be a sub-committee of the health committee.

An interesting solution to the problem has been found in Glasgow, where the chief officer of the Welfare Department is also chief administrative officer in the Health Department.

In sum, those with the widest knowledge would be likely to be least dogmatic in this field. Much depends on personalities. But in view of the foregoing considerations and the fact that the great majority of authorities in the country have decided that it is desirable to have a welfare committee and a chief welfare officer, it would seem an unhappy outcome if the likely initial arrangement in the new London Boroughs were crystallised into permanency.

II. THE TRANSITIONAL PERIOD

The foregoing pages sketch a possible reorganisation of the welfare services as it might emerge in the new London Boroughs, and suggests that such reorganisation could be expected to achieve better co-ordination and a more economic use of manpower, a more intimate and local service, and hence more efficiency.

Opponents of the Government's proposals may say that pretty pictures of what the service might be like at some future date are all very well, but what is to happen meanwhile? They may suggest, like Brutus, that between the enacting of this dreadful thing and the first motion, all the interim is like a phantasma or a hideous dream. It is in fact not easy to exaggerate the complexity and difficulty of the problems of the transitional period. Reorganisations of local government functions have taken place before, but they have generally resulted in the digesting of smaller authorities by larger ones. The operation in this case is of a very different order, the larger bodies being the subject of amputation and dissection, resulting in the death of two of them. This disappearance of the counties of London and Middlesex, the two largest local authorities in the country, poses two formidable questions. Since the administrative reputation of both is high, and London in particular has provided certain unique services in the welfare field, can the transfer be effected without loss, or at any rate without such loss as will outweigh any compensating gain? Can an immediately satisfactory method of transfer be devised in view of the fact that, whereas in outer London any initial scheme can be modified by arrangement between the new outer boroughs and Surrey, Kent, Essex and Hertfordshire, which will still be in existence, this will not be possible in inner London, since the counties of London and Middlesex will have gone.

These questions can be considered, in relation to the welfare services, from three aspects—transfer of field staff and records, transfer of residential and special establishments, and the preservation of certain special facilities and institutions now provided by the L.C.C.

The transfer of field staff and records is in principle the least daunting prospect, though in practice the volume and complexity of work involved is formidable. Both in London and Middlesex these are already dencentralised, into nine divisions and 10 areas respectively, between which and the new London Boroughs (13, excluding the City, in London, eight in Middlesex), there is a fair measure of correspondence. In London north of the river, three divisions would coincide with three new boroughs, except that Holborn would go from one to another, and the other two divisions would each be divided into two. Rather more adjustment would be required in south London, where four divisions become five boroughs, generally by the transfer of one borough. In Middlesex, three existing areas correspond with three of the new boroughs (3). Two correspond with the addition, from other areas, of one existing borough each (2); two areas combine to form one new London Borough, less two municipal boroughs (1) and the remaining three areas would have to be split up among two new London Boroughs (2). As already mentioned, London has only three welfare divisions to be split up among the 13 new London Boroughs.

The reallocation of residential and special establishments is a much more formidable proposition, but it is worth emphasising once again that the difficult transitional period in prospect can fairly be regarded as a stage in the achievement of much needed reform. Existing establishments are ill-distributed, and many are of a size repugnant to modern conceptions of welfare.

It is obvious that most, if not all of them, have residents in them originating (so far as their origins can be traced) from more than one new London Borough. It is equally obvious that no establishment can be satisfactorily controlled by more than one authority, and that therefore each individual establishment will have to be allocated to one or other of the new authorities. On what basis is this allocation to be made? Local ties must be given some consideration, as indeed they are at present to some extent, which will help, but it is certain that with present maldistribution, and the location of so many homes, especially for children, outside London, every borough will initially be managing establishments outside its boundaries. Probable demand must be taken into account. So far as possible, different types of homes must be evenly distributed among the boroughs. Special establishments of which the number is limited must be managed by individual authorities on behalf of others, as is done now in respect of remand homes and approved schools.

But what body is to be responsible for making these extremely tricky allocations, and for ensuring that a common plan be evolved and adhered to?

There has long been established by the existing Metropolitan Boroughs a liaison organisation known as the Standing Joint Committee, with its own staff and a notable record of achievement. It is to be hoped that at an early date the new London Boroughs will establish their own Standing Joint Committee, with the existing Metropolitan staff as nucleus, and that it, in turn, will set up a Welfare Sub-Committee. This is the body which might undertake this allocation, and if, as seems probable, it were found desirable, initially at least, to operate a bed bureau like that in connection with the hospitals, it might also perform that function.

It could also undertake other valuable functions, including responsibility for preserving those special facilities and institutions which have been pioneered by the L.C.C., and advising on the provision and financial responsibility of services and establishments calling for a wider catchment area than a single borough. Establishments like Woodberry Down Health Centre and the special units for difficult families maintained by the Welfare Committee would doubtless be administered by the individual boroughs in which they are located, but the Standing Joint Committee might circulate information and maintain interest in them with a view to their being paralleled elsewhere if and when practicable and desirable.

Mention has been made in previous chapters of the Central Consultative Committees on the Aged, the Physically Handicapped, the Homeless Poor and the Unmarried Mother set up by the London County Council, each of which has served as a forum for discussion between the official and voluntary bodies concerned with those particular categories of the needy. The Standing Joint Committee could maintain these consultative committees in being, at the same time enlarging their scope to cover the whole Greater London area, and so increasing their effectiveness.

It could also develop a co-ordinated policy for dealings between the new London Boroughs and voluntary organisations, which are among those who regard the reorganisation of London Government with most apprehension. They have in the past had to deal with up to nine local authorities in the London area, each with its own policies, standards and demands in the way of forms, etc. To be faced with the problem of dealing with 32 instead is more than disturbing. Moreover, the County of London in particular has shown a markedly liberal attitude to voluntary organisations, who feel that the smaller authorities might vary from the less generous to the indifferent and the positively mean. The Standing Joint Committee could evolve a policy and practice which would leave the voluntary organisations in at least as favourable a position as they now occupy.

Indeed, it is to be hoped that the voluntary organisations might be stimulated to greater co-operation among themselves in order to strengthen their position *vis-à-vis* the local authorities. It is a frequent weakness among these organisations, especially those dealing with the physically handicapped, that they tend to regard others working in the same field, whether voluntarily or officially, as rivals rather than colleagues. Yet there is now a pretty general recognition of desirability of both statutory and charitable effort in the welfare field, and a co-ordinating body of the kind suggested, operating over a field the size of Greater London might secure smoother working and more effective results as between the voluntary bodies themselves, and as between them and the local authorities. The City Parochial Foundation, one of the great charitable trusts, which operates precisely over the Greater London area, which illustrates in its history the value of the concentration of resources, and which already has close relations with most of the charitable and statutory bodies working in the area, would seem well-equipped to take the lead in such a move.

It will be remarked that no suggestion is made that the Greater London Council should play any part in the re-organisation of the welfare services during the transitional period. If, as is contended here, the London and Middlesex County Councils are too large for the effective performance of these services, *a fortiori* so is the Greater London Council. It is essential that the final responsibility should rest with the new boroughs, but equally essential that the closest co-operation should exist between them in the initial stages, and very desirable that it should continue.

III. CONCLUSION

The foregoing pages suggest that the main defect of the welfare services is their fragmentation and lack of co-ordination, and that the proposed new London Boroughs might be expected to achieve a more effective integration, not least because better communications appear to exist in smaller authorities like the county boroughs than in the counties. But they also show that the counties and London in particular, have achieved notable developments in particular services. Such developments can nearly always be traced back to the initiative and imagination of an individual, and it is contended that large authorities are more likely to attract this type of individual, because they can offer greater scope and higher salaries. This may well be true in a number of spheres of local government activity, but is doubtful in the social services. Social workers are notoriously ill-paid and enter the profession rather because of the intrinsic interest of the work than the financial inducement offered. Higher paid posts in large authorities tend to become purely administrative, and outstanding workers in this field often remain with smaller authorities, not because of lack of ambition but because of ambition fulfilled.

The larger authorities in the London area, because of their resources, are also among those making the most active contribution towards the recruitment and training of welfare staffs on the lines laid down by the Younghusband Committee, and the prospective disappearance of London and Middlesex gives rise to serious anxiety on these grounds also.

But Coventry, whose population of 276,000 is about the average of the proposed new London Boroughs, has also been cited* as having made an outstanding contribution in this field. The new London Boroughs will therefore have the resources to continue and develop the training schemes which London and Middlesex have pioneered, and every inducement to do so in their own interest, either individually or in association. They should, moreover, be able to provide a more comprehensive and integrated training if their Welfare Departments are organised on the lines suggested.

It is impossible to be dogmatic about the ideal size for a welfare authority. It has already been pointed out that for some specialist services a catchment area with a population of 166-341,000 is too small. On the other hand it could be regarded

* D. V. Donnison. Health Welfare and Democracy in Greater London. Greater London Papers No. 5.

as too large for such a widely used service as that of the home helps, and there are many indications that the larger the authority, the less closely adapted to personal and local needs its service may become. The provision of the welfare services is, however, only one of the functions performed by a local authority: for other functions other criteria may call for larger or smaller populations, and the eventual decision on size must be a matter of compromise. The compromise reached on the London Boroughs is a workable one in respect of the welfare services, but above all it would give the opportunity for their planned co-ordination in contrast with the fragmentation, improvisation, overlapping and competition which characterise their administration in Greater London today.

APPENDIX

Councils, Committees and Concern for the Old in South West Essex

I. FOREWORD

ANY study relating to old people in South West Essex must take into account the widely differing proportions of them in the different boroughs. According to the 1951 Census, these ranged from 122 per 1,000 in Leyton to 49 per 1,000 in Dagenham, and the boroughs fall fairly naturally into two groups: those with over 100 per 1,000 (Leyton, Wanstead and Woodford, Walthamstow, East Ham and Ilford) and those with under 85 (Hornchurch, Chingford, Romford, Barking, Chigwell and Dagenham), with West Ham (92) occupying an intermediate position. (see Table A.)

The 1961 Census will certainly show an increased proportion all round, but it is pretty sure that it will also show a general levelling up, the boroughs with the lower proportions in 1951 showing the greatest increases and vice versa.

II. OLD PEOPLE'S WELFARE COMMITTEES

While O.P.W.Cs. in the County of London appear to conform to some sort of pattern, those in South West Essex could hardly be more diverse. Thus in each of 22 Metropolitan Boroughs there are from one to six persons engaged whole time on Old People's Welfare, and in four others at least one engaged mainly on such work.* In all the 12 Essex Boroughs there are only two whole time workers employed. One county borough (East Ham) has no O.P.W.C. at all. Two of the other 11 authorities in effect have two, the others one each, but their expenditures range from over £40,000 a year (Barking) to under £200 (Wanstead and Woodford), (see Table B),

* K. M. Slack, Councils Committees & Concern for the Old, Codicote Press. Table IV p. 142.

whereas the range in the County of London is between £8,500 and £450.

In five boroughs (Barking, Dagenham, Leyton, Romford and Walthamstow), the Committee is virtually run by the Council (i.e., Chairman, Secretary and Treasurer are all Councillors or Borough Officials) (see Table C), but in Leyton and Romford there are two committees. In Leyton, the O.P.W.C. so called, which is affiliated to the Essex O.P.W.C., is a subcommittee of the local Council of Social Service. But there is also a "Borough of Leyton Welfare Services Committee" (of which Aldermen are Chairman and Secretary, and the Borough Treasurer, Treasurer) which provides a meals on wheels service, and has provided a chiropody service with the aid of a grant from the Essex O.P.W.C. Relations between the two bodies are quite harmonious, and the O.P.W.C. in fact runs a luncheon club for old people in the basement of the Town Hall. It was suggested to me that the organisers of the C.S.S. were ageing without the likelihood of replacement, hence the Council's action in setting up the second committee.

In Romford there is an O.P.W.C. (affiliated to the Essex O.P.W.C.) and an Old People's Committee, run by the Borough Council. Here I was told by the Chairman of the O.P.W.C. that until about five years ago they held the field, but that then the Borough Council called a meeting proposing re-organisation, as a result of which the majority of the members of the committee would be councillors. This the Town Clerk denies, and the present constitution of the O.P.C. does not bear it out, but it is clear that the O.P.W.C. disliked the idea of a greater degree of Borough Council control. There is no councillor on the O.P.W.C. but there is a representation of the O.P.W.C. on the O.P.C. The operations of the O.P.W.C., which has an income of about £200 a year, are confined to chiropody (£110), transport to Clubs (£60) and equipment of Clubs (£10) plus a certain amount of assisting of handicapped old people. Things are further complicated in Romford by the fact that there are two other independent organisations dealing with old people—the War Memorial Old Folk's Club, and the Society of Pensioners' Pals. The War Memorial Club, open daily with over 1,000 members (no subscription) is financed to the amount of about a thousand pounds a year by the Romford Carnival Week of which the Chairman of the O.P.W.C. is Chairman, and the Mayor, President. Pensioners' Pals arrange a visiting service for old people, and are represented on both the O.P.C. and the O.P.W.C.

All the O.P.W.Cs. preserve their façade of voluntary effort, and some of them the reality. Thus while in some boroughs most of the work is done by council staff there are executive committees, usually meeting every two months, on which some voluntary as well as official bodies are represented, and there is always an annual meeting to which representatives of virtually all local voluntary organisations are invited, and in some boroughs the great bulk of such work as is done, is performed voluntarily.

The amount of voluntaryism in the council controlled committees varies considerably, however. In Barking there is virtually none, a leading member of the Council having declared that all necessary services for citizens should be carried out by public authorities. Barking has practically no contact with the Essex O.P.W.C., nor has Walthamstow. On the other hand in Dagenham relations are close, and the Deputy Clerk there (who actually handles the O.P.W.C. work, although the Clerk is nominal secretary) is the vice-chairman of the Essex O.P.W.C.

In three boroughs (the M.O.H. of Ilford is Chairman in a personal rather than an official capacity) the borough is not in control of the committee.

In West Ham the secretary is the Warden of the Canning Town Settlement. She has no staff other than that of the Settlement, and minimal financial support from the local council. In the result the efforts of the committee for old people are limited, and virtually confined to the area covered by the Settlement.

In Ilford the secretary is the secretary of the local Council of Social Service. Here substantial support is given by the council, and the O.P.W.C. has done some pioneering in boarding out (rather unsuccessful) night attendant service (now taken over by the County Public Health Department) and day attendant service (as distinct from home helps) on a very small scale.

In Hornchurch the O.P.W.C. is a wholly voluntary *ad hoc* body, and an extremely active one. It has the support of the local council, but it is noteworthy that it has much the largest voluntary income in the whole area (largely derived from a local fête).

In the three other boroughs there is a mixed set-up, the chairman being a councillor and the secretary the Town Clerk, but the treasurer an outside volunteer.

In Chingford they have something resembling a federal constitution. The borough is divided into nine areas, each with an old people's association centred on a club, and the executive of the O.P.W.C. is made up of 31 members of whom 27 are members of the associations, three from each. Most of the work for the old people is carried out by the individual associations. The O.P.W.C. supports the associations, and the council the O.P.W.C. by a general grant of £100, by a grant towards meals on wheels, and by providing the administration. To quote the report of the executive to the annual general meeting: "It has not been the function of the executive to interfere in any way with the old people's organisations already so efficiently operating within the borough, but to co-ordinate and assist in old people's welfare generally and to further the work carried out by the local old people's associations of which we are justly proud."

The set-up in Chigwell and Wanstead and Woodford is somewhat similar, though the role of the clubs as local associations is not so clearly defined.

Much the least effective service to old people is given in East Ham and West Ham. In East Ham as has been said, there is no O.P.W.C. A meals voucher service has fairly recently been introduced, a limited amount of visiting and holiday provision is done by the Welfare Services Department and there is a number of old people's clubs but there is a lack of provision generally and no co-ordination of effort except in so far as the majority of the clubs are run by the W.V.S.

The situation in West Ham is peculiar. There has existed there since the war a "Veterans Club," run by a former member of the Council, which provided a club, a restaurant and a meals on wheels service, and which has been heavily subsidised by the Council in the past. Officers of the council say that they have the greatest difficulty in getting any accounts or information from this body, and as a result it is likely that support will be withheld in the future. Meanwhile an O.P.W.C. has been established (in 1956) of which the secretary is the warden of the Canning Town Settlement. This gets very little support from the council (£25 a year plus a share of the allocation of £24 to the Settlement from the Sunday Cinema Fund) and the warden has no staff beyond that of the Settlement. Consequently the amount of work it can do is minimal. There is a large number of O.P. Clubs, quite unco-ordinated, and the W.V.S. is now providing a growing service of meals on wheels, but the situation regarding co-ordination and provision generally is little better than in East Ham.

It seems that administrative factors lie at the back of these deficiencies. East and West Ham are geographically part of Essex, but administratively, as county boroughs, independent of the county council. The Essex O.P.W.C. receives generous support from the county council for services provided within the administrative county. East and West Ham are therefore outside its scope for financial assistance. They are similarly outside the scope of the (County of) London Council of Social Service and its standing O.P.W.C. The solution pending the reorganisation of London Government would appear to be for the National Old People's Welfare Council to occupy itself with the situation.

III. MEALS SERVICES

All the boroughs in the area provide meals for old people in one form or another—meals on wheels, luncheon clubs or meals vouchers.

Meals on Wheels—(see Table D). Certainly the most important of these methods of provision and one of the most valuable of all services to old people is meals on wheels, which is a method of bringing help to those who have an established need.

Only two boroughs do not provide meals on wheels—East Ham and Hornchurch, and Hornchurch has some provision under consideration.

All the boroughs having the service have at least three deliveries a week, but Romford, which has four, divides these between two areas, so that the old people receive only two meals a week.

Walthamstow is the only borough with a five day a week service. Most of those receiving meals there get them five days a week, but because of the waiting list, they may initially receive them less often, and it is interesting to note that although this is a council controlled O.P.W.C. and there are paid drivers "the meals are delivered by voluntary helpers, and it is a remarkable tribute to all concerned in this service that on no day has the service failed to deliver the required number of meals, and at all times the work of delivering the meals has been voluntary". W.V.S. is associated with the service here, but does not run it, as in most of the other boroughs.

The Leyton Medical Officer of Health has recently reported to the Welfare Services Committee, urging that a five day service should be provided there also. Here again the service is council controlled with paid drivers, and here again "voluntary helpers serve on each van". The M.O.H. said: "At present the domestic help service attempts to provide a home help to cook meals on the days the service is not operating, and on other days to those (30) on the waiting list for meals on wheels. It is of course only possible to meet a small fraction of this community demand."

In this borough all cases for meals on wheels are sifted by the Health Department, and one of the domestic help organisers is responsible for providing the van drivers with a list of cases each week.

The close relationship which exists, or should exist, between the meals on wheels and domestic help services is also stressed in the following extract from the annual report of the M.O.H. for Walthamstow.

"The provision of 'Meals on Wheels' as a five day service has made it easier to plan the Domestic Help Service, allowing a greater number of bed-ridden cases to have the comfort of an early visit from the domestic for fire and breakfast, knowing that the mid-day meal will also be received."

It is interesting to contrast the situation in Barking with that of Walthamstow.

	Barking			*Walthamstow*
	No. of meals vouchers used (*see p.* 141)	*No. of persons receiving meals on wheels*	*No. of meals on wheels served*	*No. of meals on wheels served*
1951–52	79,043	63	—	12,506
1955–56	407,595	63	12,345	19,122
1960–61	965,694	57	11,893	34,633

In the annual report of the Barking O.P.W.C., 1954-55 it was stated "The number of people who are receiving 'Meals on Wheels' remains about the same as in previous years, although if the scheme were more widely known there would be no doubt of an increase."

The two boroughs giving the most comprehensive Meals on Wheels service (Walthamstow and Leyton) which however still leaves an unsatisfied need, are both in the group with the higher proportion of old people. Nevertheless, the figures in Table D suggest there is scope for expansion in all the other boroughs, especially by way of more frequent deliveries.

In nearly all these boroughs Meals on Wheels is run by W.V.S., who sometimes have difficulty in finding sufficient volunteers to drive the vans and deliver the meals, quite apart from expanding the service.

It seems that other boroughs might well follow the example of Walthamstow and Leyton and employ paid drivers. The meals, as in those boroughs, could still be delivered by volunteers, and so retain the personal contacts which are so important a part of this service.

Communal meals. Apart from a small scheme in Leyton, luncheon clubs are to be found in only two of the boroughs, Dagenham and Romford. The bulk of the meals in Dagenham are provided in two civic restaurants with paid staff (17,000 meals a year) by W.V.S. (12,000) and by a local Settlement (10,000): two other voluntary groups also run smaller centres.

The luncheon centres in Romford are a comparatively recent venture, two having been opened in September 1961, and a third in November. Others are planned. The meals are provided and served by a caterer, but a volunteer is in attendance to take payments. Two of the centres are in halls owned by the council, the third in a restaurant, and they do not serve as and are not linked with clubs.

The merit of communal meals clubs is that while those who take them are not necessarily the most needy, they are certain of receiving a good hot meal, and do make social contacts. The centres are well patronised when provided and would undoubtedly be welcomed in other boroughs.

Meals vouchers. Meals vouchers are the least satisfactory method of providing meals for old people. Since they are made available to all old age pensioners on the production of a book, it is the forthcoming in all senses of the word who get them and only the mobile who can use them, though a wide choice of eating places is given. Moreover, they are obviously open to abuse with the result that the recipient may not get the intended hot meal—one is everywhere being told stories of their being sold, used to treat grandchildren to ices, several being blued together to provide a good blow-out and so on.

Only three boroughs make use of them yet they constitute considerably the greatest element in the total amount spent by the boroughs on old people in the area as a whole—something in the order of £60,000 out of £80,000 a year.

They cost Barking £36,213 10s. 6d. in the year ending 31st December, 1961, out of a total expenditure on the Old People's Welfare Committee of £40,140 4s. 6d. That figure omits the cost of the voucher books, £760 7s. 0d., and of the wages of the issuing clerks, another £521 14s. 1d.

The council quite frankly regard the vouchers as a means of supplementing the old age pension. In 1958-59, when the total number of vouchers issued was 783,722, the number using them was 4,567. At 9d. a voucher, this equals an average annual addition to the pension of £6 9s. od. It will be more now if the increase to 965,694 vouchers in 1960-61 represents an increase in the number of vouchers per person, as well as the number of persons.

The answers to the objections (a) that it is the forthcoming who get the vouchers and (b) that the voucher system is open to abuse are (a) endeavours are made by the party organisations in the wards to see that all old age pensioners get the vouchers. In this one organisation is said to be more efficient than the other. (b) If a well-fed pensioner wants to use his voucher for some other purpose than getting a meal or wants to give his grandchild a treat, why shouldn't he? The object of the voucher is to contribute to the well being and happiness of the pensioner.

This matter of Barking's meals vouchers is a peculiarly cogent example of the need to put first things first in the field of old people's welfare. The state having made provision for the old, adequate or inadequate, by means of pensions and National Assistance, is it the function of local authorities and voluntary organisations to do something for all who have attained an arithmetical number of years, or is it to seek out those upon whom old age has inflicted need, sorrow, sickness or any other adversity, and relieve their distress?

IV. CLUBS

The *quantity* of clubs in the area, and number of days a week on which they meet are given in Table E. It will be noted that the great majority meet once a week only, generally in the afternoon.

Of the quality it would be impossible to give an adequate picture without a detailed investigation. Obviously it varies greatly. It is noteworthy for example that in Chingford virtually all the work of the O.P.W.C. (visiting, holidays, etc.), is done through the Clubs, which must therefore play an important part in the lives of their members.

One question which cannot be answered is how far old people run their own clubs, not so much financially as administratively, and in how many they are solely at the receiving end.

Nearly all are self-supporting or voluntarily supported. Only one borough, Walthamstow, makes a substantial financial contribution "in addition to paying various charges for use of premises, the Executive Committee made a number of contributions to assist with repairs and decorations" to a total of £1,320 for their eight clubs. It is to be noted that all of these meet on four or five days a week, and in this respect Walthamstow is better served than most boroughs.

Certainly finance is one of the obstacles to a more frequent meeting of clubs elsewhere.

Thus the largest and most flourishing all-day club in the area, the Romford War Memorial Club, with over 1,000 members, costs over £1,000 a year to run, in spite of the fact that all the officers are volunteers, and rotas of volunteers take care of the kitchen providing light refreshments. There is however, no subscription, though there is an "appreciation" box and a sick fund box, and members pay 2d. a time for billiards "at their own request." The £1,000 is raised by the annual Romford Carnival.

The two all-day clubs in Chingford are also run voluntarily—one in a hall by St. John's Ambulance Brigade, one in a community centre by a rota of volunteers. The O.P.W.C. pays the rent of the latter, and also pays for newspapers in both and makes a small contribution towards refreshments.

In Dagenham the daily Veteran's Club is for men only and meets in the local Central Hall. In addition Dagenham has six day centres, Hornchurch five, and Barking two, the Dagenham centres being mostly sited in conjunction with the luncheon clubs.*

With regard to the Hornchurch daily centres (or clubs) it is said "centres are popular for three days a week (mostly afternoon attendances). For the remaining two days attendances would be very small owing to local clubs being in session."

The Barking centres are purpose built by the council, smallish but attractive. Refreshments only are provided, not meals. They are open daily except Sunday from 10 a.m. to 9 p.m. and a paid warden is in charge. Voluntary help is given in the provision of teas, and Rotary and Toc H undertook the lay-out and maintenance of the garden at the first centre opened. The second cost £5,835 to provide (site £1,110, building £4,175, furniture and fitting £550) and each costs about £700 a year to run.

The normal distinction between a club and a centre is that there is no formal membership of the latter—patrons drop in as they feel inclined, as they would to a public park or library. The daily centres are certainly appreciated where provided, but once again it is for the mobile that they make provision and they only give to the old people (or rather those of pensionable age), demanding nothing from them.

* One of them, just opened and in charge of a paid warden, was formerly the Territorial Army Headquarters, and was bought by the Council under the Physical Training and Recreation Act.

V. HOME VISITING

(See Table E.) Home visiting will normally be understood to be social contact with old people in their own homes, but it can perhaps reasonably be extended to include the ascertainment of the old people who need and welcome such contact.

To many this ascertainment seems the most urgent and valuable of all the services that can be undertaken by an old people's welfare committee. The mobile and extrovert can to a large extent look after themselves, though it is for them that most is done. The housebound and the introvert are those who suffer most from the disabilities of old age and so frequently have least done for them, simply because they escape notice.

A minority of boroughs do nothing in this matter, and it is significant to note that two of them, Barking and Walthamstow, council controlled O.P.W.Cs., very active in some directions, have no contact with the Essex O.P.W.C., the secretary of which said she had had no dealings with them for 10 years. One is inclined to suggest they may thus tend to become closed circles, immune to the penetration of informed and progressive opinion on the subject of old people's welfare.

On the other hand it is encouraging to find that in a number of boroughs much is being done.

In Chingford each of the Old People's Associations already has a visiting service, but in addition the Executive has resolved to appoint an *ad hoc* committee, consisting of representatives of local doctors, ministers of religion, W.V.S., National Assistance Board officers, health visitors and the officers of the Executive to consider all aspects of "the problem of loneliness of old people, and the inception of a scheme whereby specially qualified persons could make periodical calls on lonely old people who were unable or unwilling to avail themselves of the facilities offered by the various old people's organisations in the borough as represented on this committee".*

In Wanstead and Woodford, which also has a well organised visiting scheme "a new scheme has been introduced whereby an approach was made to elderly persons six weeks after the death of a partner. The honorary secretary noted the names of suitable persons after receiving the record of deaths in the area. The approaches made so far had been very well received and the scheme appeared to be a success."†

* Report to Annual General Meeting, Nov. 1961.

† Report to Annual General Meeting, Oct. 1960.

In Chigwell the setting up of panels of friendly visitors is stated to be among the O.P.W.C.'s first aims, and a list of old people throughout the district was being compiled at the end of 1961. The visiting was to be under the supervision of the W.V.S., who undertook the initial visit to each of the old people.

In Hornchurch "when we say that the visiting of old people is one of the basic aims of the Association we mean so much more than is contained in the jargon of that expression. From the contact between other visitors and elderly folk who are confined more or less to their own homes for the greater part of their time, springs much of our knowledge and of the needs of old people generally, their outlook and their sensitivities. The now regular meetings of the visiting committee mean that there is a constant exchange of information and experience. A recent venture was the meeting of all visitors working in the service. This too proved valuable and we shall try to repeat such meetings."* During the year 1960-61, 40 visitors made 4,404 visits to 548 old persons.

In Romford there is a registered charity (The Society of Pensioners' Pals) which undertakes visitation independently of the O.P.W.C. and the O.P.C., though it is represented on both. This has three branches, the chairman and secretaries of which meet with the officers of the Society every two months.

In the words of the Secretary "each member undertakes to visit a pensioner at least once a month, and in practice the period between the visits is usually very much reduced as the friendship between the pensioner and the visiting member grows. We are often able to assist pensioners with the minor difficulties of their lives and we are fortunate in having a member who has radio as his hobby and as we are frequently presented with rather elderly sets we usually have a few which we can give to pensioners whose own are beyond repair.

To do all this and to provide birthday presents, Christmas presents, Easter parcels and the occasional party, the three branches raise their own funds by such means as jumble sales, socials, dances, whist drives, bazaars, etc.

We visit 126 pensioners at least once a month (see above) and have a total of 53 members . . . we are in serious need of new members."

In two boroughs only, Dagenham and Ilford, are there paid whole time old people's visitors, both being members of the council staff, and not employees of the O.P.W.C.

In Dagenham the visitor is in the Clerk's Department, under the

* Hornchurch O.P.W.C., 12th Annual Report, July 1961.

direction of the Deputy Town Clerk, who, as deputy honorary secretary, gives oversight to the work of the O.P.W.C. Visiting is regarded as "one of the most important parts of the service . . . a visit demands something more than a few minutes conversation: it involves listening to a problem or problems and then setting about finding solutions" and it is not left to the whole time visitor alone.

"Many club leaders, voluntary and statutory organisations and doctors maintain close liaison with the welfare officer. She is able to visit only the most needy cases, and there is still a large gap that can only be fitted by the aid of a voluntary visiting service undertaking calls on lonely and housebound old people. A move towards such a service has begun. The W.V.S. have undertaken the task, with the O.P.W.C.'s support and the aid of some volunteers, of searching out such persons and starting a small visiting service on some of the privately owned estates. In addition, the L.C.C., who are gathering some statistics from their tenants, have undertaken to collaborate in supplying information."*

On the borough council's housing estates, groups of old people are covered by the Committee's contact service. The contacts are paid a small weekly sum (5/- a week, "to give them official status") and their duties are defined as follows:—

"i. Between the hours of 9.00 a.m. and 5.15 p.m. to contact the O.P. Welfare Officer in the event of an emergency (e.g., serious illness or injury, fire, burglary, burst pipes, etc.), also to take any necessary immediate action (e.g., telephone doctor, relatives, police, fire brigade, etc.), as the situation demands.

"ii. During the hours other than those stated above, to act for, or assist old people during emergencies and report as soon as possible to the O.P. Welfare Officer.

"In 1959-60 this service cost £64—a modest expenditure for the cover provided."*

Ilford is the other borough with a whole time Old People's visitor. Here she is in the department of the M.O.H., who incidentally is the chairman of the local O.P.W.C. It is interesting to note that both these visitors are trained social workers (see "Council Committees and Concern for the Old", p. 125).

The Ilford "Welfare Officer for Old Folk" was invited by the M.O.H. to contribute an account of her work to his 1960 annual report. Since this account gives an illuminating picture of the scope of an appointment of this kind, I quote from it at some length.

"During the year I paid 684 visits to old persons and their

* Dagenham "Welfare of the Elderly" 1959–60 (11th Annual Report).

relatives and interviewed 336 at the Public Health offices or at some place other than their own homes.

I dealt with 541 cases, 357 being new cases. The average age ... is now over 80, many being over 85 and a few over 90. Many of those over 85 are astoundingly alert and active.

Housing problems (including consideration for priority on medical grounds) numbered 69.

Persons visited while listed for, and awaiting admission to hospital, or within a few days of discharge, numbered 124. These were cases dealt with in close co-operation with the consultant physician at the local geriatric unit. The purpose of their visits was to act as liaison with Domestic Help Organisers, W.V.S., N.A.B., etc., in arranging services to ease the patient and those ministering to him during the wait between being listed for admission and entering hospital, which probably averaged two weeks. Loan of sick room equipment was arranged; laundry service for the incontinent organised; second-hand linen sometimes begged to help those whose supply was inadequate; marking tapes stitched on for those who, unable to see, or too handicapped to do it for themselves, had no one else available to do it for them; letters written to relatives and posted, club leaders informed of members' illnesses, and a multitude of other seemingly minor, but to the sick person, quite important services rendered.

Visits to these 124 following discharge were to check that domiciliary services were re-operating and adequate; and that such things as walking aids and wheel chairs, supplied for use at home were in fact working and satisfactory. Arrangements were made for alterations to the height of beds and chairs in some cases, and advice given how to take precaution against accidents. Many persons were put in touch with voluntary visitors and many neighbours seen and told how to contact me in an emergency. I leave my home and office address and telephone number with each old person whom I visit.

Thirty-seven cases were of persons whose relatives were finding it difficult to cope. . . . It was possible to find some means of easing most of these problems. Old persons will often respond to an appeal or command from a stranger far more readily than to a similar one from a close relative. To many a person of 80, a daughter of 60 is still 'a child' from whom a perfectly reasonable request is 'audacity'. The same request from a district nurse, or social worker may be immediately responded to without the slightest fuss. This is something not always appreciated by doctor and nurse. They cannot understand why relatives ask for help in bathing or blanket bathing those who to them are perfectly amenable and mobile old persons. They do not always appreciate that they see the old person's best behaviour, but that just as a child may hold himself rigid and say 'shan't wash' to his mother, so do some old people go stiff and say 'I am cleaner than you are' to a daughter.

Families asking for a holiday for the old person, or care for the old person while the family had a holiday, numbered 43. . . . All the 43 applicants were accommodated in some way. . . .

The remaining 268 cases I have called 'miscellaneous'. To me they present a strong argument in justification of the employment of a social case worker in addition to health visitors. Their problems were as many and varied as the pebbles on a beach. Among them were 40 referred to the County Welfare Department for Part III accommodation. In a few instances I felt such an application to be in the very best interests of the applicant, but in many I felt very sorry indeed that no alternative could be found. These could have lived happily and independently in old persons' flatlets and would have been delighted to do so, but the waiting list for flatlets is still very long and many of these persons were not eligible. The waiting list for the welfare homes and hostels is also long, an average wait being about 18 months to two years.

A large number of my interviews were with persons who were awaiting a flat or welfare accommodation. The waiting period is frequently a time of great strain. . . . Some persons who have longed for the day when a flat would be allocated to them panic when the dream becomes a reality. Their furniture is too large, their curtains the wrong size, they have no cash to pay the removal men, or they are too frail and crippled to go about getting estimates for moving, buying curtain material, etc., and they worry. I have been able to help several by undertaking some of the running about for them. The housing assistants also give help in this way.

Arguments with neighbours are a common cause of distress among the aged, and I have been the channel through which oil has been poured on troubled waters on many occasions. Untidy gardens and overgrown hedges are irritations that lead to an exchange of rude words and bleak looks. The Frenford Boys' Club have some kindly young members who undertake to do a little gardening without charge, and through their generosity some of these misunderstandings were swept away with the grass and hedge cuttings. Other causes for quarrels have been muddy shoes, overflowing basins, electric fires left on all night, thoughtless gossip and deafness which led to suspicion. So often, I have no remedy to offer, but I listen and a ready listener may be all the old person needs. . . .

Loneliness is not so prevalent as the press and popular opinion would have us believe. Where it is a problem I try to find a friendly visitor from some unofficial source, but organising a visiting service is a full time job fraught with difficulties, and I know the pitfalls but lack the time and opportunity to take steps to avoid them. Therefore any introductions sometimes fail lamentably.

Living alone and being lonely are two entirely different situations. Many persons living with their families are lonely because they are shut in a little back room while the remainder of the family are enjoying themselves elsewhere. . . . The state of being alone caused by death or other unavoidable partings is bearable; this being rejected by one's family is the bitter pill, the true loneliness. No voluntary visitor can take the place of a neglectful daughter who lives close at hand yet never calls, or the son who retires to bed without saying 'goodnight'. I continue to feel that the very best place for the majority of elderly persons is a home of their own with domiciliary services to help them maintain it and to supply their personal needs.

The Domestic Help Service is the most practical of all services, but is still understaffed. When helps are ill or on holiday those whom they serve often have to do without help until they return, and there may be a time lag of a week between the application and commencement of service. Some who seek the service because of a temporary disability are fully recovered again before service can be provided. Helps have to be moved frequently, and continuity of relationship, which is an important factor with old people, is lost.

Another problem frequently brought to me is the question of whether the old person should apply for National Assistance. Many who are fully entitled to it are very reluctant to apply. It may be an old lady wishes to apply but her husband objects. Sometimes children feel that it is a disgrace and offer to subsidise their parents from their own pockets. This is very commendable, but for various obvious reasons it may not be wise, and the offer can cause grave misgivings in the minds of the old persons."

I have given these extracts at considerable length, because they give a vivid and succinct account of what old people's welfare really involves. Moreover, they give some indication of the wide field for voluntary effort, even when a whole-time paid worker is in action.

VI. ASSISTED HOLIDAYS

In six of the 12 boroughs (see Table G) there is no general scheme of assisted holidays for old people. In two others there are very limited schemes for individuals, but no holiday parties.

The most extensive scheme is that of the O.P.C. in Romford, where 1,002 old people were sent to holiday camps (charge £3 10s.) and guest houses, the old people paying £2 2s. 6d. at the camps and 35/- at the guest houses.

Next came Dagenham, sending away 530 under group schemes and eight disabled old people individually.

The interesting feature here is that in this council controlled O.P.W.C., where the council is Labour, a means test is applied.

"The fact that", in the previous year, "some 195 applicants were unsuccessful whilst others had a holiday under the Committee's arrangements as well as one or more privately arranged, led the committee to review conditions under which holidays were granted. In the earlier years most of the fewer number of applicants were known personally to the welfare officer, and there was little need to regulate applications by prescribed conditions. With growth, these conditions no longer obtain and it became apparent that the continuance of arrangements on an informal basis was producing unfairness and discontent.

After a thorough review of the whole scheme the committee therefore decided:—

(a) To adjust the standard charge to 50 per cent of the cost—£2 5s. od. which includes transport—still a very cheap holiday!

(b) To exclude from the holiday scheme old people who are able to provide themselves with a holiday. For this purpose visits to relatives are not regarded as holidays and special consideration is given to cases where the sons and daughters pay for an extra week's holiday for old people living with them so that they themselves can get away with their children.

(c) To introduce a scale of payment based on applicants' income.

SCALE

	Group Couples	*Single Persons*
1.	Net income £8 plus	Net income £6 plus; not eligible

2. Net income £6–£8 Net income £4–£6; full cost (i.e., twice standard charge)

3. Net income £4 15s.–£6 Net income £2 19s.–£4 (standard charge, i.e., 45/-)

Remissions from standard charge are the amount by which the net income falls short of the lower figures mentioned in 3.

'Net income' means income less rent.

Discretion may be exercised in borderline and exceptional cases.*

An interesting and useful by-product of the scheme was that some three dozen old people who disclosed incomes below the lowest range in the scale were found to be not receiving National Assistance to which they were entitled, and were assisted by the welfare officer to obtain an adjustment." (See also p. 149.)

Hornchurch also mention "isolated instances" of old people, ineligible by reason either of income, or of already having had a holiday, being sent on an assisted holiday, and the efforts made to preclude this possibility.

It will be noticed that some of those who joined the holiday parties paid full cost, and in Hornchurch also a party was made up of full cost payers.

Here again one is left with the impression that it is the least infirm who profit most from these schemes. Only Dagenham is on record as having made special provision for disabled old people. This is not to suggest that assisted holidays do not give a great deal of happiness, not only to the old people themselves but also sometimes to relatives who are thus enabled to take a more carefree holiday themselves.

VII. CHIROPODY, INCONTINENT LAUNDRY, NIGHT ATTENDANCE

These are now all undertaken by the County Health Department. In the case of chiropody, this is a recent development and the service is still in a transitional stage in some areas. Hitherto, chiropody formed one of the chief activities of all the O.P.W.Cs., a grant of £3,000 being made by the county council to the Essex O.P.W.C. which distributed it over the local O.P.W.Cs. (see Table C). In fact some of the O.P.W.Cs. in the survey area continue to operate the service, the county council having often had difficulty in engaging an adequate number of chiropodists. At the same time a proposed reduction of grant caused a good deal of trouble with

* Dagenham "Welfare of the Elderly" 1959–60.

individual committees. Thus Hornchurch report "at the moment there is one County Clinic in the centre of Hornchurch, in addition to our seven cited throughout the area. If the withdrawal of this grant does not coincide with a full and comprehensive service being set up by the council then it will be the old people who will suffer". [The Essex O.P.W.C. report that "the executive committee have held firmly to their point that no old person who could receive needed treatment under the old scheme should have to forgo it because of the change. It is hoped that, with patience, this result will be achieved."]

Similar difficulties have arisen at Chingford and at Romford, where chiropody is provided by the O.P.W.C. not the O.P.C. On the other hand, a smooth transition is taking place in Ilford (where the M.O.H. is chairman of the O.P.W.C.) and in Chingford it was reported that in October 1961 that two-thirds of the service had now gone over to the county, and the rest was expected to go within two months.

In Barking the service was transferred on 10th July—a simple operation, since the Borough Council had previously operated the service, nominally through the O.P.W.C., and claimed the full cost back from the county direct.

Dagenham, Leyton and Walthamstow, already had borough foot clinics in operation. These are all boroughs which have their own health committees and where the county area M.O.H. is also borough M.O.H.; and some of them had foot clinics before 1948, as also had Barking.

In West Ham, a chiropody service is run by the O.P.W.C., the county borough council paying the chiropodist, and the Canning Town Settlement rent at the rate of 10/- a session for seven sessions a week, plus 10/- a week for administration. The Settlement also runs a domiciliary service, not subsidised by the council.

In East Ham there is no service, although the M.O.H. has urged its provision, as well as other services for the old, and has been turned down on financial grounds. It is significant in this connection that East Ham has the highest rate in South West Essex, and the lowest yield from a 1d. rate.

The local health departments provide both incontinent laundry and night attendance services. Neither is very extensively used. The maximum use of the former was Ilford—82 users in a year, 17,000 articles washed. From 2,500 to 11,000 articles a year were dealt with in other boroughs.

The Essex County Council decided to provide a night attendance service in the county from 1st December, 1960, after a pioneer scheme in the North Eastern Health Area. I was generally told it was very little used (Romford for example had had no case in the current year) but it undoubtedly meets a great need when it is called upon.

The Ilford O.P.W.C. (which claims to have pioneered the night attendance service in this country) is now pioneering a day attendance scheme, i.e., for cases needing more daily attention than can be provided by a home help. This was originally intended for keeping, e.g., mentally disorientated cases, out of hospital, but has developed into a service for the dying, e.g., of cancer. The local geriatrician considers it should rather be used in cases of emergency. Calls for its use are very spasmodic. The one or two volunteers on call for the service are paid 12/- a day (usually 9.00 a.m. to 5.00 p.m. i.e., when relatives etc., are at work).

VIII. MISCELLANEOUS SERVICES

(See Table H.) It is not claimed this table is complete but it gives some indication of the provision made in the various boroughs.

One additional service is a boarding out experiment operated by Ilford. A boarding out officer has been employed there for two years, but the scheme cannot be regarded as successful, as only four cases have been successfully boarded out. Funds are running out, and the experiment is likely to be discontinued, as it is felt the area is unsuitable—there is too much whole-time and part-time work available for women.

By way of contrast, it may be mentioned here that a very successful boarding out scheme is being operated by the North Kent Council of Social Service.

Since the scheme started in 1956, 222 old people have been boarded out (60 men and 162 women). Sixty-four new cases were boarded in the year ending 31st, October 1962, out of 198 applications, about a third of which came from welfare officers, and the rest from C.A.B.s., almoners, relatives, housing departments, etc. The ages of those boarded were:—

60 — 70 — 58
70 — 80 — 99
80 — 90 — 60
Over 90 — 5

A good proportion were suffering from arthritis or heart conditions, some from senility, others from nervous troubles, bronchitis, failing sight and incontinence. Twenty-five per cent were said to be in need of some care, and 5 per cent a great deal of care. Ninety-eight are still boarded out—of the remaining 124, 18 have died, 17 have been admitted to Part III accommodation, and the remainder have returned to relatives or gone to hospitals, etc.

The rates of maintenance at which cases were placed were

Rate	Cases
at £3 10s. 0d. and under	43
£3 10s. 0d.—£4	53
£4 0s. 0d.—£4 10s.	49
Over £4 10s. 0d.	65

Two thirds of the cases were provided with furnished accommodation and full board, but various other arrangements were made with part furnishing or none, and without board. In 11 cases the boarded out cases acted as paid companion, and in 12 shared expenses with the hostess.

The scheme is operated by the secretary of the North Kent C.S.S. with the assistance of six carefully selected part-time boarding officers, working about eight hours a week and paid 5/- an hour plus 6d. mileage cost for use of car, and is financed by annual grants of £1,000 from the Kent C.C. and a total of £650 from the local borough and district councils.

TABLE A.—BOROUGHS IN SOUTH-WEST ESSEX—RATES, POPULATION, AND PROPORTION OF AGED

Borough	*Area Acres*	*Rateable Value* 1.4.60	*Adjusted Rate* 60–61	*Est. Product of* 1*d. rate per head of population* 1960 £	*Resident Population* 1961	*Numbers per* 1,000 *over* 65 1951
East Ham C.B.	3,324	1,473,180	25/9	12·84	104,359	104
West Ham C.B.	4,689	2,665,785	23/0	15·47	157,186	92
Barking	3,877	1,362,727	23/0	17·98	72,282	70
Chigwell U.D.	8,971	938,896	19/8	14·83	61,001	70
Chingford	2,868	747,788	21/0	15·73	45,777	81
Dagenham	6,554	1,665,235	22/0	14·29	108,363	49
Hornchurch U.D.	19,768	1,777,060	22/5	13·66	128,127	82
Ilford	8,425	3,071,178	21/11	16·56	178,210	102
Leyton	2,594	1,366,766	22/8	13·52	93,857	122
Romford	9,342	1,606,340	22/8	13·48	114,579	72
Walthamstow	4,342	1,772,246	23/8	15·12	108,788	106
Wanstead & Woodford	3,842	1,059,915	21/4	16·54	61,259	120

TABLE B.—FINANCE OF O.P.W.Cs. IN SOUTH WEST ESSEX

	Grant from Borough Council	*Subscriptions Donations, etc.*	*Grant from Essex O.P.W.C.*	*Grant from E.C.C.*	*Other Income*
East Ham C.B.	—	—	—	—	—
West Ham C.B.	£35 £208 (chiropody)	—	—		—
Barking	£40,140	—	£91·4	£417 (chiropody)	£50 from Mayors' Fund £75 insurance claim
Chigwell U.D.	£242 (meals on wheels)	See "Other income"	£161 (chiropody) £62 (transport)		Only voluntary income is from Alexandra Rose Day & "Help Yourself" Draw N.C.S.S. £10
Chingford	£100 £80 (meals on wheels)	£45 10s.	£18 £194 (chiropody)		
Dagenham	£7,298	£70 10s. 6d.	£50 £14 (chiropody)		See also footnote to Table H
Hornchurch U.D.	£500 (holidays)	£711	£243 (chiropody)		
Ilford	£1,250 (admin. of meals vouchers)	—	£109 (chiropody)		
Leyton	£3,133	£12	£26 (chiropody)		
Romford O.P.C. }	£9,480				
Romford O.P.W.C. }	Nil		£104 (chiropody)		
Walthamstow	£4,750	£13			
Wanstead & Woodford	£500	£58 10s. (also £90 for Xmas parcels)	£68 (chiropody)		

TABLE C—STAFFING OF O.P.W.Cs. IN SOUTH WEST ESSEX

	Chairman	*Secretary*	*Treasurer*	*Membership* *C'cil*	*C'tee*	*Borough Council staff on O.P. full time*	*Part time admin work done in Council offices*
East Ham C.B.	—	—	—	—	—	—	—
West Ham C.B.	V	V	V	—	16(2)	—	No. O.P.W.C. is run from the Canning Town Settlement
Barking	B.C.	B.O.	B.O.	85(12)	—	2 part time clerks on meals vouchers 2 whole time wardens of day centres	Yes
Chigwell U.D.	B.C.	B.O.	V	42(3)	—	—	Yes
Chingford	B.C.	B.O.	V	59(3)	31(3)	—	Yes
Dagenham	B.C.	B.O.	B.O.	80(14)	14(7)	1 O.P. visitor	Yes
Hornchurch U.D.	V	V	V	69(5)	—	—	No
Ilford	B.O. (M.O.H.)	V (ISSA)	V (ISSA)	—	14(2)	1 O.P. visitor	No O.P.W.C. is a sub-c'tee of the I.S.S.A. (Ilford Social Services Association)
Leyton	B.C.	B.C.	B.O.	60	15(5)	—	Yes
{Romford O.P.W.C.	V	V	V	—	16(0)	—	Yes
{Romford O.P.C.	B.C.	B.O.	B.O.	—	24(6)	—	No
Walthamstow	B.C.	B.O.	B.O.	47(6)	22(6)	—	Yes
Wanstead and Woodford	B.C.	B.O. (M.O.H.)	V	50(10)	12(4)	—	Yes

B.C.= Member of Borough Council. B.O.=Official of Borough Council. V=Voluntary worker.
Figures in brackets indicate No. of Council representatives.

Table D —Meals Services for Old People in South West Essex

Total Expenditure by Boro' Council on Meals £	Borough	Agent	Deliveries per week	No. receiving	Meals served p.a.	Charge to O.P.	Open 5 days	3–4	1–2	Total Meals served	Remarks
		Domiciliary Mobile Meals					*Lunch Clubs etc.*				
1,500*	East Ham C.B.	W.V.S.	—	—	—	—	—	—	—	60,000 vouchers used a year	Meals vouchers @ 6d issued by Welfare Service Dept. through W.V.S. about 600 participants.
2,004	West Ham C.B.	W.V.S. Veterans Club (See text)	3	72	7,500	10d.	— —	? Veterans Club	—		A lunch club is being built by Canning Town Settlement.
38,869	Barking	O.P.W.C.	4—5	57	11,893	9d.	—	—	—	965,694 vouchers used a year	Meals vouchers issued @ 9d. about 5,000 participants.
242	Chigwell U.D.	W.V.S.	3	40	5,500	1/–	—	—	—		—
80	Chingford	W.V.S.	3	40	5,000	1/–	—	—	—		—
5,312	Dagenham	W.V.S.	4	93	19,000	8d.	4	2	—	53,874 inc. meals on wheels.	O.P.s pay 10d. at lunch clubs
—	Hornchurch U.D.	—	—	—	—	—	—	—	—		Meals service under consideration.
18,000*	Ilford	W.V.S. & I.S.S.A.	4	64	13,000	10d.	—	—	—	487,341 vouchers used a year	Meals vouchers issued @ 1/– 3,487 participants.
2,558	Leyton	O.P.W.C.	3	150	20,000	10d.	—	—	1		C.S.S. provides a meal for about 60 twice a week @ 6d. in the Town Hall.
6,500*	Romford	W.V.S.	4†	62	6,000	8d.	3	—	—	70,000 a year excl. meals on wheels	270 luncheon club meals a day. Other centres planned. O.Ps. pay 10d.
3,775*	Walthamstow	O.P.W.C.	5	150	34,633	1/–	—	—	—		—
250	Wanstead and Woodford	W.V.S.	3	35	3,000	1/–	—	—	—		—

*Estimates. †Two areas, each with two deliveries a week.

TABLE E—HOME VISITING OF OLD PEOPLE IN SOUTH WEST ESSEX

	Voluntary	*By B.C. Officer*	
East Ham C.B.	No	No	A certain amount done by the Welfare Department.
West Ham C.B.	Yes	No	Canning Town Settlement visits about 60 old people.
—			
Barking	No	No	—
Chigwell U.D.	Not Yet	No	Visiting list now being compiled. Visitation by W.V.S.
Chingford	Yes	No	Each Old People's Association has a visiting service. *Ad hoc* Committee considering how best to ascertain and approach isolates.
Dagenham	Yes	Yes	Council visitor makes about 1,200 visits a year. Voluntary visiting service being organised. Borough also has "contact scheme" on its estates. (See p. 2).
Hornchurch U.D.	Yes	No	"One of the basic activities of the Association", 40 visitors made 4,404 visits to 548 old people in 1960.
Ilford	Yes	Yes	684 visits paid in 1960 by B.C. Officer. About 4 voluntary visitors.
Leyton	Yes	No	A very limited amount of visiting by the Leyton C.S.S.
Romford	Yes	No	Society of " Pensioner's Pals". 53 visitors to 726 pensioners. Limited amount of visiting of handicapped old people by O.P.W.C.
Walthamstow	No	No	—
Wanstead and Woodford	Yes	No	Visiting sub-committee of O.P.W.C. very active. Special efforts made to find old people living alone.

TABLE F—CLUBS AND CENTRES FOR OLD PEOPLE IN SOUTH WEST ESSEX

	Number of clubs open 1 day week	*Open more than 1 day*	*Open daily*	*Day Centres*	*Remarks*
East Ham C.B.	9	—	1 (W.V.S.)	—	Six clubs run by W.V.S. Four others
West Ham C.B.	?45	—	—	—	Day Centre planned by Canning Town Settlement.
Barking	8	1 (4 days)	—	2	
Chigwell U.D.	7 (including 2 once a fortnight and 1 once a month)	1 (3)	—		
Chingford	6*	2 (2)*	2*		
Dagenham	1 (including 1 once a fortnight)	2 (2)	1	6	Day centres all provide a midday meal. A further one is planned.
Hornchurch U.D.	11	3 (3)† / 2 (4)†	—	(5)†	
Ilford	16	—	—	—	All-day centre under consideration.
Leyton	5	2	—	—	
Romford	8	—	1	—	£1,000 a year raised by Romford Carnival for War Memorial (daily) club.
Walthamstow	—	8 (all 4 or 5 days)	—	—	
Wanstead and Woodford	7	—	—	—	

*Known as "Old People's Associations," not clubs. †These are strictly speaking day centres with no fixed membership.

TABLE G—ASSISTED HOLIDAYS FOR OLD PEOPLE IN SOUTH WEST ESSEX

Facilities provided by O.P.W.C.'s

	Org. Hol. Parties	*Charge per week*	*Participants per annum*	*Help with individual arrangements*	*No. helped*	*Other Facilities*
East Ham G.B.	No	—	—	—	—	Limited service through Welfare Dept.
West Ham C.B.	No	—	—	—	—	Canning Town Settlement sends away 20—30 a year.
Barking	No	—	—	—	—	
Chigwell U.D.	Yes	59 @ £3 10s. 10 fares paid 19 full cost	87	—	—	Holidays organised through clubs.
Chingford	No	—	—	Yes	10 (with £5 each)	
Dagenham	Yes	on a scale see pp. 150-1 Minimum 45/–	530	Yes	8 disabled O.P.	
Hornchurch U.D.	Yes	30/–	165	—	—	Additional party arranged for those able to pay full cost.
Ilford	No	—	—	Yes	6 (at a cost of about 2 gns. week)	Through Women's Holiday Fund.
Leyton	No	—	—	—	—	Circulate information only.
Romford (O.P.C.)	Yes	42/6d. camp 35/– guesthouse	1,002	—	—	
Walthamstow	No	—	—	—	—	Outings arranged by O.P.W.C. and individual clubs.
Wanstead and Woodford	No	—	—	—	—	

TABLE H—MISCELLANEOUS SERVICES FOR OLD PEOPLE IN SOUTH WEST ESSEX

	1	2	3	4	5	6	7	8	9	10	11	12	13	14	15	16	17	18	19	20	21	22	23	24	25	*Remarks*
East Ham C.B.	x					x										x		x		x				x	x	These services available through Welfare Department.
West Ham C.B.	x	x	x								x		x	x		x	x			x	x					Services mainly provided through Canning Town Settlement.
Barking	x	x	x	x					x				x	x	x		x	x			x	x		x		
Chigwell U.D.	x	x	x	x					x	x	x		x	x	x	x				x	x	x				
Chingford	x		x						x	x			x							x	x	x				Cheap bowls.
Dagenham		*A	x	x					x					x							x	x				Clinic for elderly.
Hornchurch U.D.	x	x	x	x					x	x			x	x	x	x					x		x			
Ilford		x		x					x				x	x	x									x		Mayor's Christmas Fund.
Leyton		x	x		x				x	x			x			x				x		x		x		
Romford		*B				x			x	x	x	x	x				x			x	x					Fishing in lake.
Walthamstow																										
Wanstead and Woodford		x		x		x			x		x			x												Mayor's Christmas Fund.

*A Dagenham Col. 2. As a result of the Mayor's Christmas Appeal, and a contribution from the Dagenham United Charities, 2,202 old people 68—79 received a P.O. for 7/–, 405 80 and over, 10/– P.O.

*B 1,741 food parcels (£1,025) and £124 towards Christmas parties granted by O.P.C.

KEY TO TABLE H

1. Bedding and/or Clothing, provision of
2. Christmas parcel distribution
3. Christmas dinners or parties
4. Church Service
5. Clothes mending service
6. Domestic help and decorating
7. Domiciliary dental service
8. Employment scheme
9. Entertainments
10. Exhibitions of Old People's work or Talent Competitions
11. Fuel, assistance with
12. Hairdressing facilities, i.e. domiciliary washing or shaving
13. Harvest Festival gifts, distribution of
14. Home nursing requisites, loan or provision of
15. King George VI Social Service Scheme Courses
16. Library, mobile or special facilities
17. Ophthalmic Service, assistance with
18. Safety precautions, e.g. gas meter lowering arranged
19. Shoe repairs service
20. S.O.S. Cards
21. Summer Outings for the active
22. Summer Outings for the housebound
23. Welfare foods, at reduced prices
24. Wireless sets, provision or repairs, etc.
25. Special facilities at public baths

Index

For Product Safety Concerns and Information please contact our EU representative GPSR@taylorandfrancis.com
Taylor & Francis Verlag GmbH, Kaufingerstraße 24, 80331 München, Germany

www.ingramcontent.com/pod-product-compliance
Lightning Source LLC
LaVergne TN
LVHW021129300926
849160LV00010B/934
9780415864732